ROLLS-ROYCE SILVER SHADOW

BENTLEY T-SERIES

CAMARGUE & CORNICHE

Other Veloce publications -
SpeedPro Series
How to Give Your MGB V-8 Power
by Roger Williams
How to Build a V8 Engine Short Block for High Performance
by Des Hammill
How to Build & Power Tune Weber DCOE & Dellorto DHLA Carburetors
by Des Hammill
How to Power Tune the MGB 4-Cylinder Engine
by Peter Burgess
How to Power Tune the MG Midget & Austin-Healey Sprite
by Daniel Stapleton
How to Power Tune Alfa Romeo Twin Cam Engines
by Jim Kartalamakis

General
Alfa Romeo Owner's Bible
by Pat Braden
Alfa Romeo Modello 8C 2300
by Angela Cherrett
Alfa Romeo Giulia Coupe GT & GTA
by John Tipler
Bubblecars & Microcars Colour Family Album
by Andrea & David Sparrow
Bugatti 46/50 - The Big Bugattis
by Barrie Price
Bugatti 57 - The Last French Bugatti
by Barrie Price
Chrysler 300 - America's Original Muscle Car
by Robert Ackerson
Citroen 2CV Colour Family Album
by Andrea & David Sparrow
Cobra - The Real Thing!
by Trevor Legate
Completely Morgan: Three-Wheelers 1910-1952
by Ken Hill
Completely Morgan: Four-Wheelers 1936-1968
by Ken Hill
Completely Morgan: Four-Wheelers from 1968
by Ken Hill

Daimler SP250 'Dart'
by Brian Long
Fiat & Abarth 124 Spider & Coupe
by John Tipler
Fiat & Abarth 500 & 600
by Malcolm Bobbitt
Lola T70
by John Starkey
Making MGs
by John Price Williams
Mazda MX5/Miata Enthusiast's Workshop Manual
by Rod Grainger & Pete Shoemark
Mini Cooper - The Real Thing!
by John Tipler
Morris Minor, The Secret Life of the
by Karen Pender
Motorcycling in the '50s
by Jeff Clew
Nuvolari: When Nuvolari Raced ...
by Valerio Moretti
Porsche 356
by Brian Long
Porsche 911 R, RS & RSR
by John Starkey
Singer Story: Cars, Commercial Vehicles, Bicycles & Motorcycles
by Kevin Atkinson
Triumph TR6
by William Kimberley
Triumph Motorcycles & the Meriden Factor
by Hughie Hancox
Vespa Colour Family Album
by Andrea & David Sparrow
Volkswagen Karmann Ghia
by Malcolm Bobbitt
VW Beetle Colour Family Album
by Andrea & David Sparrow
VW Beetle - The Rise from the Ashes of War
by Simon Parkinson

First published in 1996 by Veloce Publishing Plc., 33, Trinity Street, Dorchester, Dorset DT1 1TT, England. Fax 01305 268864

ISBN 1 874105 64 2

© Malcolm Bobbitt and Veloce Publishing Plc 1996

Readers with ideas for automotive books, or books on other transport or related hobby subjects, are invited to write to Veloce Publishing at the above address.

British Library Cataloguing in Publication Data -
A catalogue record for this book is available from the British Library.

Typesetting (Bookman), design and page make-up all by Veloce on AppleMac.

Printed and bound in England

ROLLS-ROYCE SILVER SHADOW

BENTLEY T-SERIES

CAMARGUE & CORNICHE

MALCOLM BOBBITT

VELOCE PUBLISHING PLC
PUBLISHERS OF FINE AUTOMOTIVE BOOKS

ACKNOWLEDGEMENTS
&
INTRODUCTION

Acknowledgements

When I started to research this book I quickly realised just what a vast subject the Silver Shadow and Bentley T-Series cars was going to be. Fortunately, with assistance from Peter Baines, General Secretary of the Rolls-Royce Enthusiasts' Club, I was able to talk to Ian Rimmer, who readily agreed to check my manuscript in technical matters, and to offer advice. Apart from being Quality Engineer at Rolls-Royce, Ian is also Secretary of the Northern Section of the R-REC, as well as a member of the club's national committee. My sincere thanks to Ian - who many R-REC members recognise through his own splendid book on experimental cars - for his unfailing help.

In addition, Ian made it possible for me to consult a great many people who were involved in the development of these fine cars and allowed me to search company photographic archives. One of those Ian suggested I speak to was Martin Bourne who, until taking retirement from Rolls-Royce fairly recently, spent many years in the styling department. I cannot thank Martin enough for his wholehearted help in arranging for me to talk to the stylists, design and body engineers and test drivers, as well as senior personnel responsible for the Shadow. I am grateful to Martin for his enthusiasm and patience in answering my never-ending questions.

I am indebted to Rolls-Royce Motor Cars Ltd. for providing specification material and production details and for permission to use original draw-ings; Chris Ladley for assistance in obtaining archive photographs, many of which have not been previously published; Mark Whitaker, Co-ordinator - Marketing Communications, for permission to reproduce specific photographs; Graham Hull, Chief Stylist at Crewe for explaining styling methods, and Barry Greenwood for providing special photographic material.

Through Martin Bourne I have been able to speak and meet with many retired personnel, and I would like to thank John Hollings, Technical Director; Chief Stylist John Blatchley, who showed such foresight in shaping the Shadow, assisted by Stylist Bill Allen, who created the two-door models which later became the Corniche; Eric Langley, Body Engineering Designer who created the original full-size body layout; J (Mac) Macraith Fisher, Derek Coulson and Jock Knight, all of whom spent endless hours designing and perfecting the cars; John Astbury, who worked on engine design, and John Cooke who had to deal with many of the design regulations; John Gaskell and George Ray, both test drivers who spent innumerable hours behind the wheel of experimental cars; Dave Tod, sales representative, who took the first production T-Series to Scotland to sell the Shadow; Roger Cra'ster who, as Export Manager, negotiated all sales of vehicles to the royal family; Peter Hill, Nick Colbourne, Dave Preston and George Moseley, all of whom have added much general information.

Much assistance has been provided by the Rolls-Royce Enthusiasts' Club at Paulerspury, and I am grateful to Emma Newman, Barbara Westlake and Philip Hall. I would also like to gratefully acknowledge the help provided by the Bentley Drivers' Club at Long Crendon.

A number of specialists have also been most helpful and I am indebted to the following for providing much time and assistance: Rhoddy Harvey-Bailey at Harvey Bailey Engineering; Rob Jones, Benver Services, Crewe, John Bowling and John Bowling Jr. at Bowling-Ryan; Bill Bateman, Cumbria Classic Car Centre; Michael Hibberd, Langley, Buckinghamshire; Reg Vardy plc.; Appleyard Rippon; Dutton Forshaw, and Murray Motors.

I am also grateful to the following kind people who have provided their time and enthusiasm: Robert Vickers, Andrew Morris and Brian Drummond.

As always, I would like to thank the librarians at the National Motor Museum and, in particular, Annice Collet, Marie Tieche, Mike Budd and Terry Strange for supplying and checking information, and Jonathan Day for supplying a number of photographs.

Roger Lister has been of tremendous help in checking the manuscript for typographical errors as well as offering advice and I am grateful for his painstaking efforts.

My thanks, of course, to Rod Grainger and Judith Brooks at Veloce who suggested I write this book.

This long list of acknowledgements goes nowhere near being able to express my appreciation for the help everyone has offered. If I have omitted to mention somebody, I hope they will accept my sincere apologies.

Finally, a word of thanks to Jean, my wife, who, as always, has provided so much in the way of encouragement. She, too, has gained a passion for the Silver Shadow!

Introduction

When introduced, the Silver Shadow was claimed to be the most innovative Rolls-Royce for over 50 years. Always dignified, the marque had consistently commanded a sense of respect as a result of engineering excellence and graceful poise and appointment that epitomised good taste. With innovation came new design and build techniques that not only promised to make the car available to a wider car-buying public, but also made a substantial profit for the company. Thankfully, these changes never undermined the foundations upon which the marque had built its reputation.

Only now, looking back at motoring history, can it be appreciated just how big a technical leap forward the Silver Shadow and its Bentley derivative were for Rolls-Royce. After years of building and designing bespoke motor cars, the company - because of the demise of specialist coachuilders - was faced with a formidable challenge: accept modern motor engineering methods, build in large numbers - or die ...

Rolls-Royce, luckily, was a company rich in expertise and foresight, whose staff - people like Dr Llewellyn Smith, Ray Dorey, Harry Grylls and John Blatchley - were not afraid of change. They it was who led the company through a turbulent era when many time honoured traditions had to

be consigned to history. Fortunately, these practices, which ensured that only the best was good enough, were not forgotten. It was with some relief that when the Silver Shadow and T-Series Bentley were unveiled in 1965, complete with Pressed-Steel bodies and engineering features bristling with innovation, there remained the finesse and quality that only Rolls-Royce could achieve.

This book is not only about the cars produced over the fifteen year period from the mid-sixties to 1980, but also the people who conceived the models long before launch, those who styled them, modelled the prototypes, designed, engineered and built the definitive vehicles. This book is also about achievement and the commitment from those whose job it was to produce 'The Best Car In The World'.

Well over 36,000 Silver Shadow and T-Series cars were produced and, happily, a huge number have survived worldwide, due not only to excellent design and quality of manufacture, but also the care of owners. A Silver Shadow or its Bentley sister car, therefore, make an ideal investment as classic cars to be cherished. Although the last Silver Shadow was built something like 16 years before this book was published, its derivatives, the Corniche and Bentley Continental, have only just disappeared from the company's catalogue, a tribute indeed to the car's design and engineering.

Writing this book has been a uniquely pleasurable experience as it has brought me into contact with the very people who were responsible for the car. A particularly memorable occasion occurred just a few days before this introduction was written, when members of the Rolls-Royce styling, design and engineering departments gathered at the Crewe factory to be photographed alongside the very last Corniche built. That was a piece of history in the making which I will never forget.

Malcolm Bobbitt
Cumbria, England

Author's note
Throughout the text, reference is made to an individual's names in italics, e.g. Harry Grylls (*Gry*). This practice was very much part of the established tradition within Rolls-Royce and I have used it as an expression with historical value.

In the acknowledgements I expressed my thanks to John Hollings: sadly, John died prior to publication of this book.

6

CONTENTS

I

THE BEST CAR
IN THE
WORLD

There was no doubt what was the main attraction at the 1965 London Motor Show at Earls Court: lamps high above the array of gleaming cars seemed to shine on two stands in particular - Rolls Royce and Bentley - where a constant stream of visitors gazed in curiosity and pride at the models making their first British appearance.

Any new model Rolls-Royce or Bentley has always been something of a sensation and the Silver Shadow and

T-Series were no exceptions.The two cars had already enjoyed an enthusiastic reception at their European debut two weeks earlier at the Paris Salon, which opened on Thursday 7th October and continued until Sunday 17th October. It was now the turn of British motorists to see for themselves the latest models to earn the reputation of "best in the world".

There was some doubt about whether the Silver Shadow and T-Se-

The Silver Shadow was unveiled at the Paris Motor Show in the autumn of 1965 and had its British debut at the London Motor Show, where this picture was taken, a couple of weeks later. The Silver Shadow, of course, was the first Rolls-Royce and Bentley model to have unitary construction and Rolls-Royce took the unusual step of displaying the car without doors in order to show the build structure. (Courtesy Rolls-Royce Motor Cars Ltd.)

Two special-bodied S-Series Continental cars shared the Bentley stand with the new T-Series model at the London Motor Show. Mulliner, Park Ward's elegant coachwork is illustrated in this drawing of a Flying Spur.
(Courtesy Rolls-Royce Motor Cars Ltd.)

ries cars would be prepared in time for the Paris Motor Show, especially as there was already frenetic activity to produce the cars in time for the British launch. There is evidence that some commentators considered that the cars on show in Paris might not have been driveable as the bonnets were kept firmly locked and rumours suggested they were without engines; hardly likely considering that the models on show had been demonstrated to Rolls-Royce and Bentley agents at Crewe on 30th September and 1st October.

Whereas some manufacturers might have seized the unveiling of a new model as a chance to introduce a little *razzmatazz*, Rolls-Royce, with customary sophistication, resisted the impulse. *The Autocar* described the appearance of a new Rolls-Royce or Bentley as the event of a decade and this was especially so with the Silver Shadow and T-Series which the magazine acknowledged as being by far the most advanced and intricate cars the company had so far produced. The cars from Crewe were certainly revolu-

tionary; Rolls-Royce claimed they were the most radical new cars in 59 years. The specification was impressive and included monocoque - integrally chassied - construction, hydraulically self-levelling suspension, all-round disc braking, independent rear suspension, and much more. The resultant design was a car which was more compact in size, being lower and shorter than previous models, yet which lost nothing in the way of interior space and comfort.

Rolls-Royce and Bentley cars retained their own identities at Earls Court and were featured on separate stands. Stand 112 was graced by three Silver Shadows: one was finished in Regal Red and another in Dawn Blue, while the third was presented in Shell Grey. In spectacular contrast to the new model, and virtually dwarfing it, was displayed the majestic Phantom V Limousine which, like all Rolls-Royces and Bentleys, was built to special order. Unlike its predecessor, the Phantom 1V, of which all 18 models built had been reserved exclusively for royalty and heads of state, the Phantom V

was available for general purchase and in total 832 examples were produced.

Bentley cars adorned stand 141 and the marque which had once enjoyed the accolade of being The Silent Sports Car, was represented by two elegant T-Series, one in Caribbean Blue and another in a striking two-tone colour scheme of Sage Green over Smoke Grey. Two special-bodied S3 Continentals by Mulliner, Park Ward (who, by this time, had been absorbed by Rolls-Royce as a subsidiary company), were also displayed. Although the T-Series Bentley superseded the S-Series cars, the Continental version was still listed as a special order and the two vehicles - one a dashing white convertible with scarlet leather upholstery, and the other, a Flying Spur, finished in Shell Grey and Blue - appeared quite generous in proportion to the new model.

It could not be missed, however, that, as separate marques, Rolls-Royce and Bentley had gradually grown closer together and had been sharing designs. This was even more apparent

9

Although the Silver Shadow and Bentley T-Series had almost identical styling features, the cars nevertheless retained their own distinctive radiator shells. The difference in price between the two cars was £60.00.
(Courtesy Rolls-Royce Motor Cars Ltd.)

Rolls-Royce and Bentley motor cars were synonymous with engineering excellence. The company's aero engines powered the world's first jet airliner, the DeHaviland Comet, seen here with an R-Type Bentley Continental.
(Courtesy Rolls-Royce Motors Ltd.)

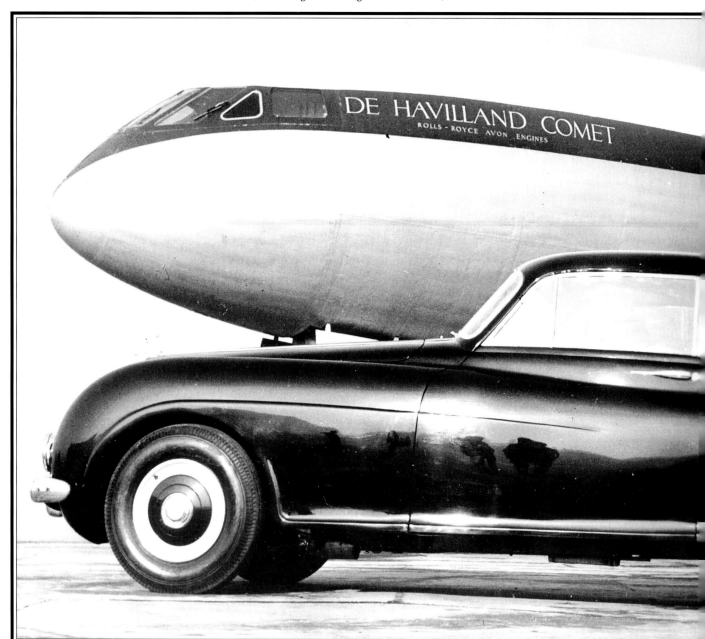

with the new generation cars and the T-Series Bentley, as noted in the pages of *Motor* and *Autocar*, had become, essentially, a badge-engineered version of the Silver Shadow. The price difference between the two cars was so slight as to be insignificant and the future of the Bentley version seemed less secure than that of the Rolls-Royce. To many customers another £60 to acquire the prestigious Rolls-Royce badge would have seemed a bargain and there was the danger that due to Rolls-Royce-preferred sales, the Bentley marque would become extinct. Whilst many would recognise and appreciate the Bentley emblem for its associations with motorsport traditions, it was the famous Spirit of Ecstasy mascot designed for Rolls-Royce by Charles Sykes in 1911 that commanded the most respect for British engineering at its best. Not only had Rolls-Royce become synonymous with motoring excellence, but the company's aero engines now powered some of the world's best known aircraft, including the D.H. Comet, the first jet airliner.

Those who gazed at the new Rolls-Royce and Bentley cars at the 1965 London Motor Show could have been forgiven for being amazed at what they saw; the beloved Silver Cloud, ten years old and carrying with grace its huge, razor-edge styled coachwork, was no more and the new car was, for Rolls-Royce, a totally new concept in design.

The Silver Shadow was Rolls-Royce's most radical car in over 50 years. This was due to its chassisless construction which meant that external dimensions could be reduced without compromising interior space or comfort. Note the difference in radiator size between the Silver Shadow on the left, and the Silver Cloud, right. R. Allwright stands between the two cars in his life. (Courtesy Rolls-Royce Enthusiasts' Club)

The Silver Shadow was sleek, its coachwork smooth, and there was a large glass area but, more than anything else, it possessed a degree of dignity and sophistication that could only be expected from a Rolls-Royce.

The ten year old Bentley S-Series and Rolls-Royce Silver Cloud design was not without its critics within the motor industry. For long enough harsh words had been uttered about the cars' handling and, if the Crewe factory's design team was affected by such criticism, it was not shown publicly, such was the team's loyalty and pride to the company and cars it produced. Instead, the design team worked diligently to produce a car making full use of the latest technology available to the motor industry.

The Silver Shadow and T-Series cars were the first Rolls-Royce and Bentley models to use a chassisless construction. This was sensational enough and yet the cars were fitted with all-round disc-braking, which ensured more than adequate stopping power, and a radical suspension system that afforded the most comfortable cosseting for driver and passengers. The cars were seen as an important departure from traditional pro-

duction methods, a brave move which coupled advancing technology with the need to be more accessible to a wider and increasingly more discerning market.

Criticism of the Silver Cloud and S-Series cars was replaced by an enthusiastic, if perhaps slightly cautious, reception for the new generation models. Some enthusiasts were more than a little reluctant to accept the company's new direction and saw it as a departure from a tradition steeped in history. The decision to proceed with a radically different design of car was not taken lightly but, ultimately, it was survival in a competitive market that was the deciding factor.

Whilst it might appear that traditional coachbuilding techniques had been discarded in favour of volume production methods, the change to unitary construction was no more than a process of evolution. Some aspects of Rolls-Royce manufacture could, understandably, never change. The company's legendary attention to detail and refinement remained as always and both Rolls-Royce and Bentley cars retained that very special quality which was second to none and of which the marques were justly proud. Ironically,

the styling of the Silver Shadow allowed a return to traditional values in respect of the famous Rolls-Royce radiator: the lower profile of the car meant that radiator height could revert to the proportions established during the pre-First World War era.

The postwar period

The changes to Rolls-Royce manufacturing methods which resulted ultimately in the Silver Shadow and T-Series cars was the culmination of many years' painstaking development. The programme can be traced to the period immediately prior to the outbreak of the Second World War and it is understandable that the decisions taken at that time were considered quite extreme. This was a time when the design of the automobile was undergoing serious change, streamlining was becoming popular and there was a move towards producing small and economical family cars, a good example of which is the Volkswagen concept and Ferdinand Porsche's endeavours to produce a car for the people. When the political situation in Europe deteriorated and the likelihood of war very real, Rolls-Royce - along with other manufacturers

12

At one time all Rolls-Royce and Bentley cars were built with separate chassis. This is the chassis of the short-lived prewar Bentley Mk V. (Courtesy Rolls-Royce Motor Cars Ltd.)

such as Rover and Austin - was forced to alter its role dramatically. Car production was put aside and manufacturing concentrated upon the war effort.

Rolls-Royce's factory at Crewe had been constructed as part of the Shadow Factory Scheme introduced in 1936 to produce aircraft components in preparation for war. Work began on the Pym's Lane site during the second half of 1938 and, within 11 months, the first Crewe-built Merlin was running. By the end of the war the plant had produced 26,000 Merlins of various marks, and 2000 Griffon engines. During the war years development of the site continued with the addition of machine and assembly shops and test rigs. Shadow Factories, which were often camouflaged for protection from aerial attacks and painted to look, from the air, like housing terraces, now provide an interesting area of industrial archeology. Incidentally, signs of the camouflage were still faintly visible in the early seventies. Once their wartime use was over many Shadow Factories were turned over to car building and, as with the Rover Company, which had adopted a Shadow Factory at Solihull as its car manufacturing plant, so Rolls-Royce concentrated motor production at Crewe.

Traditionally, Rolls-Royce produced bespoke motor cars of the highest quality which earned the acolode of best in the world. Whilst Rolls-Royce hand-crafted the chassis, bodies were prepared by the most famous coachbuilders who applied their ex-

acting standards and degree of refinement. The cars were supplied to customers who demanded only the very best and graced some of the most influential motor houses around the world. The history of Rolls-Royce and Bentley as separate entities has been well documented. Rolls-Royce continually lived up to its formidable reputation and Bentley was 'The Silent Sports Car'. In 1931 Bentley was absorbed by Rolls-Royce in a move which followed a controversial merger attempt. Napier, which had entered car manufacturing in 1900 but was better known for aero engines, saw the chance to gain control of Bentley and re-establish itself in the motor industry as a producer of quality cars. Rolls-Royce considered that Napier's engineering prowess, coupled with Bentley's automobile experience, would pose a serious threat and out-bid Napier at the last moment.

The Second World War had a dramatic affect upon Rolls-Royce as a company. Prewar, car output had never been huge and virtually half of all cars produced were exported. When car manufacturing resumed in 1946, there was a potentially huge market for motor cars, but it was left to manufacturers of the popular makes to meet that demand. Even though the cars produced were, essentially, of prewar design with only a minor face-lift, there

seemed no shortage of buyers except, of course, for luxury cars. With home sales deeply depressed in this market, it was to export sales that Rolls-Royce turned its attention.

Decisions taken when reviewing some of Rolls-Royce's manufacturing methods were the result of plans drawn up at the end of the 1930s by W.A. Robotham (*Rm*) who, at the time, was head of chassis design. Having joined Rolls-Royce as an apprentice in 1919, Robotham was appointed to the company's experimental department in 1923 and soon made a name for himself working with, and testing, chassis. His brief called for standardisation of designs which would have the consequence of lowering production costs without any loss of build quality. Some of the ideas that Robotham suggested were far-reaching which, had they been allowed to fully materialise, might have drastically affected the whole structure of the company. In 'popularising' the Rolls-Royce and Bentley marques to the extent that he had foreseen, the demise of the company as it was, and as it is now, might have been a distinct possibility. According to John Blatchley, who was responsible for the Silver Shadow's styling, Robotham was keen to forge a relationship with the Rover company. Any such plan, of course, did not materialise.

The Mk V Bentley was the first car to receive a standardised chassis design. Pictured here in London's Richmond Park is one of only 11 cars built. (Courtesy Rolls-Royce Motor Cars Ltd.)

In demonstrating the company's willingness to consider new ideas, Rolls-Royce experimented with the design of a small car known as the 'Myth', a vehicle of diminutive proportions - by Rolls-Royce standards - which weighed 22 hundredweight (1118kg) and which was equipped with an engine of just 1497cc. Decidedly problematic, the prototype car was continually beset by overheating and was dismantled after a mere 407 miles (651km) had been completed.

Standardisation of chassis design and the introduction of component interchangeability was a realistic venture first seen on the Bentley Mark V, a very short-lived model which was abandoned due to the outbreak of war and after only 11 cars had been produced. The series was not revived when, postwar, production was resumed.

For bespoke motor cars such as the Rolls-Royce and Bentley, it might have seemed somewhat unnecessary to introduce standardisation, especially when annual output of vehicles amounted to somewhere around only 1000 cars. It was, however, low volume production which demanded standardisation: with less than 100 cars a month being built it was likely that as

many as three chassis designs, all quite different, would be needed.

W.A. Robotham foresaw that the bespoke bodywork crafted by leading coachbuilders so usually associated with Rolls-Royce and Bentley would, eventually, become a practice of the past and that the company would either produce its own bodies or buy in from a volume supplier. It was accepted that such a change of direction would have to be a gradual process and the first steps towards standardisation were taken almost unnoticed.

Rolls-Royce selected a number of coachbuilders who could produce bodies specifically for standard chassis designs and it was to these that customers were directed. This presented little problem, especially as the customer had been assured that the coachwork was specifically approved by Rolls-Royce and Bentley. These companies included James Young of Bromley, Park Ward of Willesden and H.J. Mulliner of Turnham Green, Chiswick. The association between Rolls-Royce and Park Ward developed to such an extent that the latter company was absorbed by Rolls-Royce in 1939. H.J. Mulliner was taken over by Rolls-Royce in 1959 and in 1961 the two coachbuilding concerns of Park Ward and H.J. Mulliner were merged and became known as H.J. Mulliner, Park Ward Ltd. The close association with Rolls-Royce did mean that 'in-house' coachbuilding methods

After the war, Rolls-Royce resumed car production with the Bentley Mk VI. Here, a Park Ward convertible is given its final inspection at the coachbuilder's London factory.   (Courtesy Rolls-Royce Motor Cars Ltd.)

The Bentley Mk VI was also available with a standard steel body. Note the semi-razor-edge styling of this, the first of the postwar designs. (Courtesy Rolls-Royce Motor Cars Ltd.)

of the highest quality could be maintained. James Young, however, remained as a separate company and continued to supply bodies for Rolls-Royce and Bentley cars from its Bromley premises.

After the war ended the Bentley Mark VI heralded the return of car production at Rolls-Royce. The detail and design of the car was not dissimilar to that of the Mark V Bentley; the classic semi-razor-edge styling clearly having an influence on the postwar car. Standardisation had arrived at Crewe ...

Even though car production ceased during the war years, development and research on W.A. Robotham's ideas nevertheless continued. Rolls-Royce played an important role in the war effort and it was fortunate the company had the means to also carry on with experimental work and produce some prototype vehicles. The Rover company at Solihull was in a similar situation and Maurice Wilks was able to try out a number of particular designs on 'company business and time'.

Standard steel bodies

Convinced the future lay in a chassis

design capable of accepting a standard body format, Rolls-Royce went shopping for the best quality product. The company it approached was Pressed Steel of Cowley, Oxford, in January 1944. Pressed Steel, with its premises opposite the Morris factory, provided bodyshells to much of Britain's mass-produced motor industry and had the potential facilities to supply to Rolls-Royce.

Pressed Steel readily agreed to produce bodies to Rolls-Royce's own specification (the prestige was enough to secure the deal) and it was Ivan Evernden (*Ev*) and John Blatchley (*JPB*) who undertook the styling. While Evernden produced the basic engineering principles it was Blatchley who took responsibility for the detailed design.

H. Ivan F. Everndon's career with Rolls-Royce and its design department started when he joined the company in 1916 as a young man. Six years later, in 1922, he found himself a protégé of Sir Henry Royce when he was appointed to the design team, at that time based at West Wittering on the south coast of England. Extremely conscientious by nature, Ivan Evernden built a reputation for precision

(amongst the cars he designed was the R-Type Bentley Continental, one of the all-time great sporting cars). Following his retirement in 1961 he continued for a period of time in a consultant capacity and even, in a moment of meditation, designed his own headstone in the shape of a Rolls-Royce radiator shell!

John Blatchley was no stranger to Rolls-Royce either as, prior to the Second World War, he had worked for J. Gurney Nutting, coachbuilders of considerable repute. The company, established since 1919, had been building bodies on Rolls-Royce chassis since 1925. Generally favoured by Rolls-Royce in its standardisation programme, Gurney Nutting was acquired by Jack Barclay Ltd. in 1945. Whilst John Blatchley's design followed the fashion of the immediate prewar period, it also incorporated the latest styling ideas which were considered in keeping with the new Rolls-Royce and Bentley models.

John Blatchley's early career is one of immense interest. As a boy he suffered from rheumatic fever and did not have a proper schooling. He was educated at home by a private tutor and his thoughts were often not on his

studies but motor cars. His early drawings, which were sketched before he had reached his teens, were clearly advanced and show some very elegant coachwork styling. At the age of 18, John's father thought it appropriate for him to go to Cambridge University and, after some intensive tuition, he was ready to sit the entrance examination. The young Blatchley was less keen on an academic career, however, and, after sitting at a desk for 20 minutes (in which time he had not written a word), he put away his pen and left the examination room.

Instead of going to university, John enrolled at the polytechnic for motor body building, a course which lasted for 18 months. A friend of his worked for Gurney Nutting and it was through him John was able to get a job. As-signed to Gurney Nutting's Chief Stylist, A.F. McNeil, John Blatchley's talents were soon recognised and some of his first styling exercises were based upon Duesenberg and Lagonda chassis for an Indian Maharaja. At the age of 24 he was promoted to take over from McNeil on his retirement. In 1944 or 1945, John sought a position with Rolls-Royce and was offered a job in the company's aero engine division. He disliked designing aero engines, much preferring to work with motor cars, and eventually went to see W. Robotham at the car division.

Robotham welcomed him with open arms and Blatchley soon found himself working alongside Ivan Evernden.

One of John Blatchley's first assignments was to prepare some drawings for the proposed 'Junior' Bentley and a small Rolls-Royce, the Silver Ripple; these designs did not materialise but John's watercolour sketches of these cars are a delight. He was then given the task of tidying up the design of the MK VI Bentley which, at that time, was less than pretty. John re-styled it so that the door hinges were concealed. He was then responsible for the R-Type Bentley and later generation Rolls-Royce and Bentley models.

A new era began for Rolls-Royce in the spring of 1946 when the company announced its first postwar models, the Bentley Mark VI and the Silver Wraith. This was something of a frustrating period as components and materials were still in short supply after the war and production was severely affected. As a result delivery of cars did not commence until late in the year. Decisions made prewar in respect of standardisation meant that the two cars had certain similarities, although the Bentley was some 7 inches

Chief stylist John Blatchley at work shaping a wax model of the Phantom Limousine. (Courtesy John Blatchley)

The first Rolls-Royce cars to be built in postwar years were the Silver Wraith and this 4-door touring saloon featured at the 1952 London Motor Show. (Courtesy Rolls-Royce Motor Cars Ltd.)

Before and after the war Rolls-Royce attempted to design a smaller car. Although the project never material-ised, this 1940s drawing of a junior range car depicts the styling evolu-tion of postwar models. (Courtesy Rolls-Royce Motor Cars Ltd.)

(178mm) shorter. Whilst the Bentley was built at Crewe as a complete car, the Silver Wraith was constructed more in the Rolls-Royce tradition with the chassis being made to accept custom-built coachwork. It was three years later, in 1949, that greater standardi-sation was achieved when Rolls-Royce announced the Silver Dawn, which shared both the chassis and dimen-sions of the Bentley Mark VI. Whereas the Silver Wraith had been designed to accept a coachbuilt body the Silver Dawn was specifically intended to carry standardised steel coachwork.

The difficulties experienced dur-ing early postwar production were not unique to Rolls-Royce. All British manufacturers were at the mercy of a motor industry starved of raw materi-als which were often, even when avail-able, of indifferent quality. The govern-ment of the day ordered a massive drive to earn much needed foreign currency and priority was given to export sales. As a builder of luxury cars Rolls-Royce's market was ex-tremely precarious. Major business came not so much from home sales but the more lucrative markets abroad. In order to build cars with specialist coachbuilt bodies, Rolls-Royce would have had to continue to endure very low production figures, unacceptable if the company was to stay in business.

Likewise, it would have been impossi-ble for specialist coachbuilders, which were usually small businesses, to in-vest in huge outlays in order to raise production levels. Of the relatively few coachbuilding specialists only Park Ward, itself already absorbed into Rolls-Royce, was able to produce a stand-ardised body. Even so, it was still only possible to achieve a maximum pro-duction figure of 10 bodies a week.

Adoption of standard steel bodies did, as can be appreciated, attract some criticism and traditionalists feared that the unique appeal of the bespoke Bentley or Rolls-Royce might be lost forever. Conversely, the compa-ny's policy attracted a new clientele and it was evident that, ultimately, the design of the standard steel saloon was

as elegant as its forebears and the finish just as 'superior'. For Rolls-Royce, the change to utilising bought-in bodies meant establishment of a completely new industry. This enabled the company to finish the bodyshells delivered from Pressed Steel to accept-able Rolls-Royce standards and to erect production areas which accommodated the processes necessary to fit out and trim the cars. For the committed tradi-tionalist, however, which standardisa-tion could never satisfy, it was still possible to specify a custom-built ver-sion of either car that could be pre-pared by one of a number of firms to include Hooper, James Young, Freestone and Webb, H.J. Mulliner, Park Ward, J. Gurney Nutting and Abbot of Farnham. Continental

The first postwar Rolls-Royce specifically designed to carry a standard steel body was the Silver Dawn. (Courtesy Rolls-Royce Motor Cars Ltd.)

coachbuilders were also able to offer their designs for Rolls-Royce and Bentley and amongst the more favoured were Franay, Pininfarina and Graber.

The Mark VI Bentley and Rolls-Royce Silver Dawn evolved into a new generation of models with deliveries beginning in June 1952. The R-Type Bentley superseded the Mark VI but the Silver Dawn name was retained for the new Rolls-Royce. Both cars shared modifications and these included a larger boot and the option of an automatic gearbox, the latter initially supplied for export models only. There were some important chassis modifi-

cations which accounted for the enlarged bodywork and therefore the springs were made both longer and wider. Possibly one of the most elegant, charismatic and desirable of all postwar Bentleys was the R-Type Continental, the performance and looks of which successfully managed to revive

the marque's sporting tradition. Only 238 examples of this superlative car were built and, fortunately, a healthy number of these cars have survived.

The Silver Dawn and its Bentley equivalent, both with ancestry traceable to the immediate prewar era, remained in production until the early

The R-Type Bentley with its larger boot, as pictured here, superseded the Mk VI; the Rolls-Royce equivalent retained the Silver Dawn nomenclature, however. (Courtesy Rolls-Royce Motor Cars Ltd.)

The Silver Cloud and its Bentley equivalent, the S-Series, were introduced in 1955 and became popular abroad, especially in America. Here, a Bentley SI is being exported and is carefully loaded aboard the S.S. Brandager. (Courtesy Rolls-Royce Motor Cars Ltd.)

part of 1955 when two new models were unveiled, the Rolls-Royce Silver Cloud and the S-Series Bentley. Matters concerning future new models cannot be hurried at Rolls-Royce and, as the new cars were being acclaimed, in the design department attention had already turned to a successor for launch a decade ahead.

Commendation for the Silver Cloud was received particularly from America where it enjoyed considerable popularity. The Bentley marque, sadly, never received the same recognition from that side of the Atlantic, even though both cars shared an almost identical profile apart from the radiator shell and bonnet line. There was, however, a significant detail difference between the two cars: the bonnets were not interchangeable due to the shoulder radius which, on the Silver Cloud, was sharper than that of the Bentley.

John Blatchley was responsible for the styling of both cars which had gained the reputation of being particularly handsome. Some enthusiasts would argue that the Bentley was more aesthetically pleasing than the Silver Cloud and the S-Series was especially

America loved the Silver Cloud and this first-series car makes a handsome companion to a Piedmont turbo-prop airliner. The aircraft appears to be a Fokker F27, built under licence by Fairchild, which makes it an FH227. The significance of the photo. is that the aircraft was powered by Rolls-Royce Dart engines. (Courtesy Rolls-Royce Motor Cars Ltd.)

John Blatchley's evocative styling of the Silver Cloud and S-Series Bentley is clearly evident in this picture, which shows the latter car. (Courtesy Rolls-Royce Motor Cars Ltd.)

sought-after in Britain. This may be why more Bentleys were sold despite the Silver Cloud's prestigious radiator and mascot.

As was to be expected, the finish and trim on both cars was exceptional and in Standard Steel form their elegance could hardly be bettered. Specialist coachbuilders, however, did offer some exotic designs and Freestone and Webb produced a magnificent two-door fixed-head coupé, whilst H.J. Mulliner produced some outstanding convertibles, a feature often favoured by American customers.

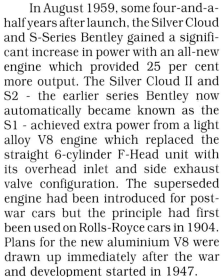

In August 1959, some four-and-a-half years after launch, the Silver Cloud and S-Series Bentley gained a significant increase in power with an all-new engine which provided 25 per cent more output. The Silver Cloud II and S2 - the earlier series Bentley now automatically became known as the S1 - achieved extra power from a light alloy V8 engine which replaced the straight 6-cylinder F-Head unit with its overhead inlet and side exhaust valve configuration. The superseded engine had been introduced for post-war cars but the principle had first been used on Rolls-Royce cars in 1904. Plans for the new aluminium V8 were drawn up immediately after the war and development started in 1947.

Exterior styling of the Silver Cloud II and S2 may have remained virtually identical to the first series cars but interior trim was modified to include facia layout and a driving position which was revised in response to the cars' top speed of over 115mph (184km/h) and slightly more athletic acceleration. As a Grand Tourer the S2 Bentley Continental boasted unequalled elegance and a feature of these cars was the reshaped bonnet which added to the car's sporting appearance. Throughout the forties and fifties, the Silver Wraith was still a true bespoke coachbuilt Rolls-Royce. Whilst the model shared a number of features with the original 1946 car, the new V8 engine eventually led to the vehicle's demise in 1959 in favour of the Phantom V, a variant based upon the Silver Cloud II chassis which offered truly

The experimental department at Crewe was constantly developing new ideas. Look carefully at the huge tail fins fitted to this S-Series Bentley, which were devised to test for crosswind stability. (Courtesy Rolls-Royce Motor Cars Ltd.)

Buckingham Palace makes a fine backdrop for this Silver Cloud II on what is a warm summer's day. (Courtesy Rolls-Royce Motor Cars Ltd.)

exceptional coachwork styling.

The final phase in Silver Cloud and S-Series model development was introduction of the Silver Cloud III and Bentley S3 towards the end of 1962. These cars are instantly recognisable by impressive frontal styling which incorporates a twin headlamp system, a styling feature popular at the time and found on a number of cars. Styling differences went a lot deeper than the headlamps, though: front wings were restyled and the radiator profile lowered along with the bonnet line; combined side lamps and indicators were fitted, as were redesigned front overriders. The interior of the cars also received a face-lift and the specification included separate front seats. Revised mechanical specifications in-cluded modified power steering and an increase in performance. Interestingly, a Silver Cloud IV and Bentley S4 had been envisaged which would have been equipped with disc brakes and independent rear suspension. This car did not materialise due to the advanced development of the Silver Shadow which incorporated these features.

A Shadow on the drawing board

Until the introduction of the Silver Shadow and Bentley T-Series, it was almost inconceivable that cars of such revered marques could be built without a separate chassis. Monocoque construction, while hardly a new innovation which involved building the chassis and body together as a complete unit, was more usually associated with the mass production methods of the more popular type of car. Builders of specialist and luxury cars, however,

Some of the coachbuilt cars were particularly elegant, including this James Young two-door coupé. (Courtesy Rolls-Royce Enthusiasts' Club)

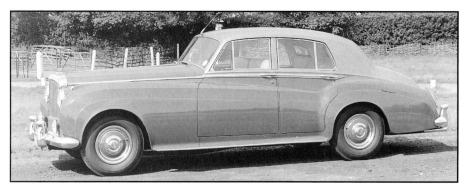

The S2 was the Bentley equivalent of the Silver Cloud II and this particular car completes what is quintessentially an English scene. (Courtesy Rolls-Royce Motor Cars Ltd.)

5933 DH

Experiments were carried out to find a suitable four-headlamp system for the Silver Cloud and S-Series cars. In this photograph, taken in the styling department, several ideas are being tried in plasticine. Note the differences between the left-hand wing line and twin headlamps and the right-hand wing, as well as the differing styles of indicator lenses. In the background is stylist Martin Bourne. (Courtesy John Blatchley)

were normally expected to use what were considered more traditional manufacturing methods.

In designing the new cars' specifi-

Twin headlamps were a feature of the final Silver Cloud series and S-Series Bentleys. The styling treatment is very evident on this Silver Cloud III which is standing outside the main entrance of the Pym's Lane factory. (Courtesy Rolls-Royce Motor Cars Ltd.)

cation, the decision to adopt monocoque construction was taken at an early stage, even before the Silver Cloud had been launched. For traditionalists the Silver Cloud and Bentley S-Series retain the importance of being the last standard production cars from Crewe to utilise a separate chassis, apart from the few Phantom models built to special order. Rolls-Royce's plans were quite specific: the new car

The concept of the Silver Shadow was the work of Harry Grylls (without whom it almost certainly would not have existed) who is shown standing between Captain Vautier (left) and John Blatchley (right). The date of the photograph is not known (possibly early fifties) but there is certainly some interest in the Silver Ghost. (Courtesy Martin Bourne)

Final painted plasticine model of the car to supersede the Silver Cloud and S-Series Bentley. (Courtesy Rolls-Royce Motor Cars Ltd.)

had to comply with future trends of an auto industry catering for an ever-increasingly car-conscious society. As traffic volume increased, the design of the motor car would change to meet the need for smaller exterior dimensions without sacrificing interior space and comfort. If the car was to be shorter in length and width than the Silver Cloud, but roomier inside and with greater luggage capacity, Rolls-Royce would have to accept that a monocoque construction was the only answer.

The move towards standardised chassis design and adoption of standard steel saloon bodies suggested that it was a matter of natural evolution for Rolls-Royce to consider integral, or unitary, construction for the new models. This was especially relevant, bearing in mind it was intended that the cars would lead production throughout the second half of the sixties, span a whole decade and more into the eighties. Development costs, which included tooling-up for the new car, were acknowledged to be exceptionally high and therefore it was expected the cars would enjoy a long production run. The change to standard steel bodies had proved entirely successful with previous models and it was anticipated the new generation of cars would receive an enthusiastic reception, especially from America which was seen as possibly the most important export market. The American motor industry had almost entirely been given over to building cars in the unitary style and by the mid-fifties Europe was belatedly

following its lead.

Even if he did not already need convincing about unitary construction methods, Ivan Evernden produced, shortly after a visit to Detroit in 1955, an important report on why a monocoque design should be adopted by Rolls-Royce. It was from America that the integral chassis and body had originated in the late 1920s when the Budd Organisation had shown some exciting prototypes to the motor industry. Amongst the manufacturers who took up Budd's ideas was Lancia in Italy as well as the French manufacturer, André Citroën. It was the latter's celebrated Traction Avant which astonished the motoring world in 1934 with its sleek design. Twenty one years later the same company produced another remarkable car, the DS19 which, together with unitary construction, featured an advanced hydraulic system that controlled not only braking, transmission and steering but suspension also.

First indications of Rolls-Royce's commitment to producing a monocoque-chassied car can be traced to February 1954 when the project of developing such a vehicle was given the code name Tibet. The design engineers at Crewe were already busy perfecting the Siam project, which was the development of the Rolls-Royce Silver Cloud and S-Series Bentley. Only after the launch of the cars had it been possible - apart from the vaguest of outlines - to give proper thought to the proposed car. It was not until 1958, when the

Silver Cloud and Bentley S-Series were fully established, that serious consideration was given to the new Rolls-Royce, and the Tibet programme was expanded to include a Bentley version, allocated the code name Burma. Originally the project was limited to a Rolls-Royce badged car but the huge costs involved encouraged senior management to reconsider their decision. By marketing the Bentley (at the time intended as an altogether smaller vehicle), an opportunity arose to widen the models' appeal. Other variants were also considered and these were given the equally exotic code names Korea, Java, Bengal, Tonga and Rangoon. Development of these projects came about as a result of an odd collaboration between Rolls-Royce and the British Motor Corporation (BMC). Interest in a sports car (again, in collaboration with BMC), which was designed to rekindle the Bentley marque's sporting heritage, brought about a brief interlude with an unlikely model which, had it been developed, would have been known as the Alpha. This, and other experimental cars, are dealt with in more detail later.

Rolls-Royce's experimental projects have a history of being allocated curious code names; throughout much of the early postwar period they were far-eastern countries. Harry Grylls (*Gry*), who was the car division's engineering director during this period, chose the names personally and they do not appear to have any particular significance other than to suggest he was a romantic with a sense of humour! Harry Grylls was appointed chief

An historic photograph as this is the only picture of the members of the Crewe design department. Taken in August 1956 when plans were afoot for the Silver Shadow, the occasion marks the retirement of chief designer, Bill Hardy. From left to right (front row): Agnes, Muriel and Thelma (surnames unknown); Harold Peak; Geo. King; Frank Tarlton; Vivian Stanbury; Bill Hardy; Reg Davies; Jack Phillips; Bert Jeal; Ralph Lever; Mavis Bonsall; Cecily Jenner; Margaret Jones. (Second row): Ted Holland; Reg Spencer; Harry Bamford; Frank Holt; Les Robinson; Eric Langley; Fritz Feller; Bill Allen; Horace Kirby; Eric Howarth; Norman Webster; Geo. Clarke; Harry Taylor; Charlie Monk; Geo. Cooper. (Third row): Cliff Evans; Arthur Horsnall; Reg Swinburne; Jim Burnham; Alan Lord; Jim Edwards; Ron Biggins; Bob Hill; Bill Condliffe; Arnold Turnaley; John Everett; Alan Tomlinson; John Gorman; Peter Pryke; Ralph Allcock. (Fourth row) : Ken Trinder; Geo. Tearle; Ernie Knibbs; Brian Carpenter; Alan Jobson; Aubrey Scragg; Gordon Linney; David Mason; Martin Bourne. (Courtesy Martin Bourne)

engineer in 1951; he had joined the company 21 years before, in 1930, when aged 21. He was well accustomed to Rolls-Royces, his father having bought one of the first cars the company produced. Always a keen driver, Harry Grylls ensured he regularly tried for himself the company's experimental cars, at every stage of development, and often, after 1953, at Oulton Park. He also made sure he knew what other manufacturers were planning and tried out their cars whenever he could.

Although the Tibet project had been on the minds of Harry Grylls and Ivan Everndon for some time, the first indication the design team had of the new car coincided with the launch of the Silver Cloud. Bill Allen (*JPB/Aln*), who joined Rolls-Royce as a stylist on August Bank Holiday Monday, 1935, remembers Harry Grylls calling the department together and, in a stunned but excited atmosphere, telling everybody the news that the Silver Cloud's eventual replacement would be an altogether radical new car.

John Blatchley, who was chief stylist under Ivan Evernden, was very enthusiastic at the prospect of designing such a car, even to the extent of losing sleep over it. When asked whether styling the car around a monocoque shell worried him, John was emphatic that the concept allowed him greater freedom from the usual constraints that a separately chassied car presented. Unitary construction, he added, provided the opportunity of lowering the passenger area to inside of the four wheelarches, so maintaining passenger comfort within a shorter frame.

As to the design of what was ultimately to become the Silver Shadow and T-Series Bentley, stylists such as Bill Allen and Martin Bourne were given a relatively free hand. It was appreciated that a monocoque car would not only be lower in profile than a separately-chassied car, but that it could support full-width frontal styling. As well as being more compact, which was considered a necessary styling feature, it had actually been possible

to achieve greater cabin space than on previous models. There were parameters on which the styling team worked, such as a 126 inch (3200mm) wheelbase, which was 3 inches (76.2mm) greater than that of the Silver Cloud, and sub-frames front and rear to support engine, gearbox, rear axle and suspension. The wheelbase of Burma, incidentally, was 119.5 inches (3035mm), 3.5 inches (88.9mm) shorter than the Silver Cloud's. The suspension system was, from the outset, designed to be nothing other than standard but it was the hydraulic self-levelling - utilised to provide an even keel under all road and load conditions - that made it different. Hydraulically self-adjusting suspension was considered a necessity not only due to the car's sheer size and weight but to appeal to the American market which was used to cars with extremely soft springing. Two power units were originally proposed for the Silver Shadow: a 6-cylinder 3990cc in-line 'F' engine for the Burma project, and the 'L' 6230cc engine for Tibet. The latter, of course,

The codename of the project intended to replace the Silver Cloud was Tibet and a full-size mock-up is shown here alongside its predecessors. On the far left is the Silver Dawn, Silver Cloud is in the centre and Tibet on the far right. It is interesting to see how the designs progressed over the years and that the Tibet, even in the early stages of development, is remarkably similar to the definitive car.   (Courtesy Jock Knight)

was utilised for 'SY', the product which resulted from the merger of the two projects.

John Blatchley, as chief stylist, was ultimately responsible for the new car's definitive outline and it was he who took the first steps to produce the shape. The body styling evolved firstly from a number of outline sketches and, once these had met with his approval, a model was crafted from wax. Supplied in large blocks, the wax had first to be cut into pieces and warmed gently in an oven before it could be used. It was difficult to control the heat of the oven and it was something of a relief once the wax achieved the correct consistency and could be applied as a whole to the modelling table. The bulk of the model was a large empty 1/2 inch (13mm) ply box, padded out by glued-on blocks of balsa wood. The final wax thickness was usually no more than about 3/8 inch (12mm) as it was very expensive.

Stylist Martin Bourne scrapes away at the modelling clay on a scale model of the Tibet. Note the distance between the wheels and the door leading edge, shorter by 6.5 scale inches (165mm) on Burma and hence Silver Shadow, the shape of which is quite evident, even though there were many changes before the definitive form emerged.
(Courtesy Martin Bourne)

The working surface of the table, which included a measuring bridge, consisted of a large panel of duralumin approximately 1/4 inch (6mm) in thickness that had a grid of 2 1/2 inch (63mm) squares - representing 10 inches (254mm) on the actual car - scribed upon it. The same method of measurement was also applied to the bridge in order to gauge correct vertical levels. The idea of the measuring

table was to produce an accurate quarter-size scale model, which was usual in the early planning of a new car. As John Blatchley carefully and painstakingly scraped away at the wax block the design department watched as his styling theme emerged.

The stylists' modelling and measuring table had been devised by Bill Allen in 1951, before the design department was installed at Pym's Lane.

The department at the time was situated at Clan Foundry, Belper, one of Rolls-Royce's many sites, and John Blatchley had just been appointed chief stylist with the immediate aim of producing a new standard saloon, the Silver Cloud.

Luckily, wax as a material is extremely forgiving and, when more was needed to fill any hollows or raise the surfaces of the model, it was easy to prepare and apply. Experiments were carried out creating a number of wax models which depicted different styling ideas. The only tools which would work the wax suitably were commercial paint scrapers and it was by using these that the stylists became skilled craftsmen. Wax is highly vulnerable to changes in temperature and, after the Christmas holiday when the factory heating had been turned off, large cracks had appeared in the model John Blatchley had been working upon. Fearing the worst, the stylists were relieved to see the cracks disappear as soon as the central heating was turned on and room temperature began to rise. While the stylists worked at sculpturing the body design, others in the department concentrated on the chassis layout, created the interior design and devised accessories such as bumpers and lamps.

The model was then drawn, quarter scale, using thousands of spot dimensions from the bridge, and scissor-cut tracing paper and card templates, the result being passed to Bert Jeal, who was in charge of the Body Office, for scaling up to full size.

Once formed, the fragile nature of the wax made the model susceptible to the most minor knocks and scratches. To prevent damage it was decided to use the example as a pattern for a mould which would allow a permanent model to be made from dental plaster. Although this proved a lot more durable than the wax it was disadvantaged by being extremely heavy and cumbersome to move. By using fibreglass as an alternative to dental plaster a satisfactory solution to the problem was eventually found, making presentation of styling proposals to directors and sales staff somewhat easier. Producing quarter-size models was also more cost effective than embarking upon expensive full-size mock-ups before definitive styling had been accepted.

Looking at a scale model and visualising a full-size car was very difficult to do. Dr (Doc) Llewellyn-Smith (*LS*), Harry Grylls and Ivan Evernden could, and it was Evernden who spent his last few weeks before retirement designing

and building an optical viewer which adjusted the perspective of the model to that of a full-sized car. It was only when 'democratic' styling committees came along, consisting of people from other departments, that full-size models were introduced to make visualisation easier.

No attempt was made, as some manufacturers had, to introduce 'see-through' quarter-scale models. The use of perspex for the windows, whilst providing an interesting three-dimensional aspect, would have made any alterations to the roof and glass areas extremely difficult to accomplish, especially as minor detail changes were constantly being tried out.

Considered just as important as external styling was the car's interior arrangement. Rolls-Royce's attention to detail and thoroughness meant that the most minor control or switch had to have the right *feel* about it and would operate in a manner commensurate with the quality of the car.

Above all else there remained in the stylists' mind concern at what the customer would make of the design. Monocoque construction meant sacrificing many old ideals, such as a flat floor in the rear compartment and, for the first time, introduction of footwells for the rear passengers. The slab-sided coachwork might have been considered featureless at the time but eventually, of course, the styling became quite acceptable.

Once a design was approved, the quarter-scale drawings were supplied to the Body Design department where the drawings were made full-size. Eric

Tibet in the experimental garage at Crewe. Note the high-mounted indicators on the front and rear wings; clearly evident is the length of the wheelbase, especially the distance between the front wheels and the base of the screen. When the Burma project got underway (which had more influence on the ultimate Silver Shadow design), the wheelbase was shortened. (Courtesy Jock Knight)

Langley (*EAL*), second in command to departmental head, Bert Jeal (*HPJ*), remembers laying out the body full-size on to 15ft by 5ft (4572mm x 1524mm) sheets of pale green aluminium, marked out with a 10 inch (254mm) grid. These formed the master layouts from which the entire prototype car was constructed. The whole process of styling, designing and building a prototype car was naturally a time consuming affair, taking approximately two years to complete.

Experiments begin

From the time the Tibet project was first mooted it was something like four-and-a-half years before an experimental car appeared in August 1958. Bearing the registration number 5 ELG, the first Tibet prototype car, which was built on chassis number 41-B, showed a surprising likeness to the definitive outline of the future Silver Shadow. In evidence was a clean and unfussy straight-through waistline. Not only was the design thoroughly modern but the car had a markedly lower profile in comparison to that of the Silver Cloud. Any link with previous models was incidental, apart from the adoption of a fashionable twin headlamp system, a styling feature later incorporated into the third series of Silver Clouds and S-Series cars which, at the time, was still

four years away from production.

Features of 41-B were prominent front indicators, built into the upper leading edge of the wing tips; ultra-slim windscreen pillars, and the oddly-shaped, wide rear quarter panel. Equally prominent were the indicators built into the top edge of the rear wings. For initial testing purposes and in order to prevent the car from being recognised as a Rolls-Royce, the prototype was without the traditional radiator and, instead, was fitted with a plain example which resembled something akin to that found on the Bentley. The car was, however, later photographed with the usual Rolls-Royce radiator in place, probably to test for wind noise. The camouflage went a stage further with the concealment of anything that might suggest the car's origin, including all brightwork. When seen against the production Silver Cloud, 41-B appeared very modern with its slab-sided bodywork and immense window area.

The car's first outing on road test showed there was much to be done to perfect the prototype: excessive noise was noted coming from the rear axle and the front brakes had an unacceptable squeal; the steering was too heavy and not nearly flexible enough, while some discomfort was experienced from vibration that stemmed from the transmission system. As for performance,

road-holding was reasonably good although spoilt, perhaps, by too much roll on cornering. Overall, road noise was found to be acceptable considering the car's unitary body shell, a feature which often took on the qualitites of a sound-box. A problem that did exist and about which customers of the standard steel saloons generally complained was the difficulty experienced in closing the doors - a complication hardly ever experienced with coachbuilt cars.

For all its less favourable aspects 41-B was nevertheless a remarkable achievement, considering that the design department at Crewe had little experience of the engineering concept demanded of it. Martin Bourne, who joined the design team in September 1955, recalls how each stage of Tibet's development was treated with a degree of both excitement and caution, the department as a whole feeling its way in what was essentially unknown territory. An overstatement, probably, but Martin considered it a wonder that the exercise was made to work as well as it did. In retrospect the project worked due to the care, attention to detail and sound engineering for which Rolls-Royce was renowned.

41-B was tested by Tony Martindale (*AFM*), Rolls-Royce's chief development engineer, whose experience with Rolls-Royce and Bentley cars dated to prewar days. The weight of the car was almost 38 hundredweight (4256lbs/1930kg), which was virtually 3 hundredweight (336lbs/152kg) less than the kerb weight of the Silver Cloud in its final version, the laden weight of the Silver Cloud III as tested by *Autocar* in August 1963 had been about 44 hundredweight (4928lbs/224kg). Considering this, 41-B's weight was little more than that of the experimental Siam project car that pre-dated the production Silver Clouds and S-Series Bentleys.

There is much evidence to suggest that the development of the Tibet project, with its numerous and radical innovations, many of which broke with tradition, placed a heavy strain upon the design team at Crewe. Bert Jeal and his team were clearly worried about dispensing with a separate chassis and building a monocoque shell to the required rigidity. The need for heavy duty sub-frames fore and aft was recognised but it was features such as all-round disc braking and the innovative self-levelling suspension system which caused much of the concern. Rolls-Royce had been working towards perfecting disc braking for some time and it had been intended that the Silver Cloud III and S3 Series cars, as well as a Series IV version which never materialised, should have been so equipped. In the event disc brakes were never used on the Silver Cloud or its Bentley equivalent and it was left to the tried and tested hydro-mechanical brakes of the Series I and II cars to serve throughout its production life.

The suspension for Tibet was initially designed around a compressed air system and, although this worked well enough while the car was in motion, it suffered from a loss of pressure once the vehicle was left standing for any length of time. Invariably the system deflated overnight and required re-pressurising from a compressed air line each morning. Air pressure was used to produce a constant ride height and the system devised was quite unlike that of Citroën's hydropneumatic or BMC's hydrolastic designs. To maintain the correct suspension level, Rolls-Royce's engineers adapted for the purpose a diesel injection pump which was different to the pully-driven belt operating a high pressure pump as found on the suspension system of Citroën's DS19. Rolls-Royce considered Citroën's practice unreliable and could possibly lead to suspension failure had the belts broken or suffered extensive wear. In order to examine Citroën's hydraulic suspension, which had first appeared as a self-levelling device on the rear axle of the 6-cylinder Traction Avant in 1954, before being developed to control the futuristic D-Series cars, Rolls-Royce purchased a DS19 and progressively dismantled it. This was common practice in the motor industry and research through Rolls-Royce's archives shows that a considerable number of cars were so evaluated, including the rotary-engined NSU RO80 and, in more recent years, the Renault Espace multi-purpose vehicle. Air suspension was not entirely new to Rolls-Royce, as the company carried out experiments with it during the war years.

In order to resolve 41-B's vibration problems the car was taken out of the experimental programme and sent to the Bump Rig for several weeks' severe testing. Work in the meantime was progressing with the second prototype car in the Tibet project and this was delivered to the test department during the first few weeks of 1959. Carrying the registration 835 FLG on chassis 42-B, this prototype was relatively short-lived; it was considerably heavier, by 150lbs (68kg), than the previous car and had been designed to utilise the same type of suspension layout which had proved as problematic as with 41-B. In less than a year 42-B had been withdrawn from the test schedule to allow the suspension to be completely modified to a hydropneumatic system. Testing resumed on the car in October 1960 but it appears that the vehicle remained troublesome as only something like 3000 miles (4800km) were recorded before the car was withdrawn less than a year later. Eventually dismantled, the car did not quite reach 20,000 miles (32,500km) in total.

A third experimental car, 43-B, which was registered 500 GMB, was delivered to the test department during the middle of 1959. This, too, suffered from a number of problems but more seriously, a series of under-bonnet fires were recorded. The cause of the fires was ultimately attributed to seizure of the oil pump drive. 43-B was heavier than its two predecessors; half a hundredweight (30kg) more than the second experimental car.

The Burma project - a wax model of the car is shown here - was originally intended as a Bentley-badged car. Ultimately, the Tibet and Burma projects were cancelled and the best features of both cars combined in what became the SY. The definitive shape of the Shadow is evident, but look at the ungainly wing line. The frontal styling is unlike any Bentley but note the peak above the windscreen, which is reminiscent of the BMC Farina cars, and which was a popular styling feature of the fifties. John Blatchley, incidentally, was required to keep the angle of the windscreen within 32 degrees, a task he found particularly irksome. (Courtesy John Blatchley)

The Burma project

Whilst development of the Tibet project had been a continuous operation, at the same time the design department at Pym's Lane was under some pressure to improve and modify existing models. The Silver Cloud II and Bentley S2 were due a face-lift and once this had been achieved and the Series III cars successfully launched, there was more time to meet the demands of the new car.

A couple of months after the first Tibet car was produced, attention was turned towards a Bentley version, intended to be fundamentally different to the Rolls-Royce. Codenamed Burma, the design was for an altogether smaller car which, whilst using unitary construction similar to that of Tibet, had a planned wheelbase of a fraction under 10 feet - 119.5 inches (3035mm). Burma was both narrower, by 2 inches (51mm), and lower by 2.5 inches (63.5mm) than Tibet. In addition it sported 15 inch wheels (381mm) whereas Tibet was fitted with 16 inch (409mm) wheels. The car was also lighter by 781lbs (354kg). The cabin space of Burma remained identical to that of Tibet and reduction in the car's overall length was achieved by removing 6.5 inches (165mm) from between the scuttle and the front wheels.

It was originally intended that the Burma car would be equipped with a smaller engine than had been envisaged for Tibet. From the outset the Bentley was designed around a new Rolls-Royce F-series 6-cylinder 4-litre engine (3990cc), unlike the Tibet which was always intended to use the L-series 6230cc V8 unit. Ultimately the Burma project abandoned the 4-litre engine in favour of the V8 as the former was not powerful enough. John Astbury, who worked on engine design at Crewe from August 1959, recalls it was a mammoth task squeezing the V8 into the confines of the Burma bodyshell. He remembers that space was so restricted, it took more time to find a location for the dipstick than to resolve almost any other problem!

When work initially started on Burma there was some question as to the design of subframe required for the car. Jock Knight, who had been involved with chassis design on Tibet, remembers sitting at the drawing board with John Blatchley and discussing the matter at some length. Eventually they devised a layout that is now familiar on the Silver Shadow.

Before prototype cars could be built, the usual quarter-sized scale models based upon outline sketches were produced in wax. Once approved, detailed drawings were created before a full-size model was produced. This was standard procedure throughout the motor industry and allowed design engineers the facility to make alterations before committing to the building of expensive prototype vehicles (a senior design engineer at Rolls-Royce once calculated that a prototype car cost 17 times more to build than a production car!) Early scale models show the Burma styling evolved through a number of ideas. The frontal appearance could, at one time, have been considered somewhat ungainly; front and rear wings changed in shape and on one model the car was seen with prominent tail fins! Headlamps were inset into a full-width radiator grille and direction indicators installed into the leading edges of the front wings. As successive design changes developed there emerged three distinct phases, each aspect of which became all the more dignified until it reached something close to the definitive model.

Burma experimental cars numbered seven in total and were built on chassis numbers 51B-57B inclusive. The first of the cars, which weighed 5cwt (254kg), experienced a number of problems including poor performance and braking, axle noise and a general rattle from the door trims as well as excessive exhaust noise. For a new car, however, there was less fouling and clunking than expected. A particularly impressive feature of 51-B was its stable roadholding, quite unlike that of the early Tibet cars. The feel of the car was made all the better by its seats which allowed a high driving position with good visibility.

Although 51-B had a profile not dissimilar to that of the Tibet cars, its frontal styling was not in the least pretty. The huge, full-width mesh grille was divided in the centre by a vertical bar and the headlamps were positioned either side in the upper corners. Indi-

Styling modifications brought about the Burma 3 with its revised frontal treatment and softer wing line. This full-size mock-up is pictured inside the experimental garage. (Courtesy Rolls-Royce Motor Cars Ltd.)

Burma 3 from the rear. The styling incorporates many of the original Burma features but the definitive Silver Shadow is very evident. The plinth surrounding the car reads 'Rolls-Royce Silver Cloud Chassis'. (Courtesy Rolls-Royce Motor Cars Ltd.)

cator and side lamps were built into the leading edge of the wings in similar fashion to the definitive Shadow. The shape of the rear quarter panels meant that it was difficult to see the rear wings from the driving seat so, to redress this situation, the rear styling was given tail fins which also housed the rear lamp assemblies. Less than 8000 miles (12,800km) were recorded on the car when it was dismantled in January 1964, the reason being that it was used extensively on the bump rig to investigate noise levels. The prolonged testing resulted in some suspension parts cracking and the car was stored from 1961 until it was scrapped in 1964.

The second Burma car, 52-B, had similar styling to 51-B but was 38kg (0.75cwt) lighter. Endurance testing was quite intensive; it was road tested by day and bump rig tested at night. At Oulton Park the car underwent steering and handling tests but, after only

26,000 miles (41,600km), it was scrapped.

53-B, the third Burma car, had a surprisingly short life. Entering service in the early months of 1960, it was involved in two accidents and was broken up in August the same year with less than 14,000 miles (22,400km) recorded. Apart from accidental damage, 53-B was a poor performer and suffered transmission and hydraulic problems.

The fourth Burma car, 54-B, which carried registration number 120 MLG, was taken to France for testing. Test driver John Gaskell remembers taking the car around Le Mans where he was photographed taking a break from the arduous test programme. John recalls that the car suffered a number of problems, especially with the hydraulic throttle, but the main complaint was that the heater failed to work to any extent and test drivers had to continuously stop the car just to be able to get

some exercise and warm themselves. After trials lasting something like 18 months (in which time 71,000 miles (113,600km) were recorded), the car was put into storage and later dismantled.

All the Burma test cars had been fitted with right-hand steering until 55-B, the fifth experimental vehicle. With left-hand steering, the car entered service at the beginning of 1961 and tests were generally conducted on the M1 motorway. John Gaskell remembers driving at high speed along the motorway and passing another Rolls-Royce or Bentley. The driver of the car recognised the vehicle that passed him at such high speed had to be a new model and quickly contacted the factory to see if he could buy one! 55-B was used to test air conditioning equipment and was also subjected to a ferocious spell on the bump rig. So severe was the testing that the body had to be rebuilt although the suspension stood up to the constant battering remarkably well. It is a matter of interest that 55-B was used as part of Rolls-Royce's and BMC's coalition, which is explained in greater detail later.

The two final cars in the Burma

Burma, built on chassis 54-B, underwent extensive trials in France. The car was also tested at Le Mans, where this picture was taken, and test driver John Gaskell takes a welcome break. In 18 months of trials, some 71,000 miles (113,600km) were recorded. (Courtesy John Gaskell)

project were 56-B and 57-B. Both cars underwent some restyling and were given a much more attractive frontal appearance. Previous cars had a pronounced peak above the windscreen but this was now removed. The cars also looked better with their wrap-around bumpers and lowered rear wing line. Test driver John Gaskell was again involved in the trials of 56-B in France and after some 20,000 miles (32,000km) the car returned to Crewe. Burma cars were, of course, fitted with the 4-litre engine but 56-B was eventually equipped with the 6230cc V8 unit in 1963. Before this, though, tests were carried out using a 6-speed gearbox and revised suspension. Tests continued with 56-B until the Silver Shadow was being put into production and assisted in that car's familiarisation programme. In January 1966, after completing 60,000 miles (96,000km), the car was withdrawn from service.

57-B was the last of the Burma cars and, although originally fitted with the 4-litre engine, was converted to a V8 between 1962 and 1963. The Burma programme was eventually scrapped, as was this last experimental car, which was used for barrier crash testing.

Whilst the Burma project was progressing, parallel work continued on Tibet. Following trials with 43-B the first Tibet phase came to an end and phase II commenced. Aptly named Tibet II, outline work on the project began as early as the closing months of 1958 but it was to be more than three years before the first development car

made its appearance. When it did, 44-B displayed many of the features from Burma and, together with its Rolls-Royce radiator, it revealed much of what was to come in the shape of the Silver Shadow.

Tibet II's frontal styling - with its twin headlamp system - was strikingly modern and aesthetically pleasing; front direction indicators were positioned lower on the wing's leading edges than previously and changes to the bonnet allowed a one-piece design which was hinged at the front. Shortly before this stage in the car's development a centrally hinged bonnet, as found on the Silver Cloud, had been contemplated. Clearly, the Burma development had considerable influence on Tibet II as the car was equipped with doors made from aluminium alloy rather than steel. The Burma project had experimented with aluminium in order to reduce the overall weight of the car and it also proved more resistant to corrosion. Adversely, aluminium was more expensive and, being a softer metal, was also more prone to damage from minor accidents. Rover, incidentally, utilised a similar material, known as Birmabright, on the P4 range of cars (60, 75, 90 and 100 models, etc.) before reverting to steel shortly before production of the series ended. The reason for the move away from Birmabright was financial, the cost of aluminium being much greater than steel.

Much experience was gained from 44-B; the suspension system was modified so that the hydropneumatic system operated as a self-levelling device on the rear axle only, while coil springs

were substituted at the front. The car, however, completed only a nominal mileage, less than 3500 miles were recorded before the car was withdrawn and eventually dismantled.

Alongside Burma, which was essentially a Bentley-inspired project, a Rolls-Royce version of the same car carried the codename Tonga. Eventually, the separate projects were cancelled and development continued in a single direction. A further reason for the decision to combine both areas of research was the huge costs of developing two quite different models. It became clear that the Bentley would not be anything other than a badge-engineered alternative to the Rolls-Royce.

An odd coalition

As Tibet and Burma became obsolete, the search for replacements to the Silver Cloud and Bentley S-Series was conducted, combining each of the car's best ideas. The new project received the code name SY. This notation came about due to Rolls-Royce's engineering department's identification codes and was retained throughout the remainder of the car's development.

During the development period of Tibet and Burma, Rolls-Royce had almost succumbed to a strange courtship with the British Motor Corporation (BMC). This could have had a devastating effect upon the Crewe company and the cars it produced, had the relationship evolved fully. BMC was looking for a place in the luxury car market, certainly something with greater appeal than already existed in

Burma 3 experimental car, built on chassis 56-B. Note how the frontal styling has changed from that of 54-B in the previous picture. (Courtesy Rolls-Royce Motor Cars Ltd.)

its top of the range models, and Rolls-Royce appeared an attractive means of achieving this. The collaboration between the two companies began in 1962 and, whilst several themes were thought of and design exercises carried out, there was only one model which came to fruition as a production vehicle - the Vanden Plas Princess R.

A hybrid of the Austin Westminster 110, the car was of handsome appearance and its features - though restyled - had origins as a Rolls-Royce project, codenamed Java. In place of BMC's 3-litre engine was the 4-litre F60 6-cylinder Rolls-Royce engine which, it will be remembered, had been originally developed to power the Burma car. Had the project evolved as intended when collaboration between the two concerns commenced, Java could have become a mass-produced 'small' Bentley. First appearing at the 1964 Motor Show the 4-litre R enjoyed only limited success, probably because the exterior image of the car suggested

an Austin rather than a Rolls-Royce. The car's price was also a disadvantage; at £1995 it was more expensive than the Rover 3-litre coupé and the Jaguar S-Type and only fractionally cheaper than the Jaguar Mark 10. Within three years the car had been discontinued.

BMC had supplied to Rolls-Royce the front-wheel-drive 1100 and 1800 full-size mock-ups for evaluation and Martin Bourne can remember being sent along to a disused test bed to carry out detail measurements of both cars, which were still very secret. What was of interest were the dimensions of the 1100's cabin which, surprisingly, offered just as much space as the Silver Cloud's! Styling exercises were carried out upon what was known as the Bengal theme and quarter scale models show a car which shared identifying features with the 1800, and which at the time was being quite

A coalition existed between Rolls-Royce and BMC during the sixties but the only model to come out of it was the Vanden Plas Princess 4-litre R. Rolls-Royce's development model was the Java, shown here with its unique frontal styling and Bentley badging. (Courtesy Rolls-Royce Motor Cars Ltd.)

seriously considered as a future Bentley. Possibly the oddest creation was a variation coded the Rangoon which utilised the Rolls-Royce radiator; on a relatively small car this feature was totally out of place. When collaboration with Rolls-Royce ended it is apparent the design eventually progressed from the BMC 1800-derived car to a

The only product of the Rolls-Royce and BMC coalition was the Vanden Plas Princess 4-litre R. The engine used for the car was the F-60 6-cylinder unit designed by Jack Phillips. (Courtesy National Motor Museum)

A further styling theme was a proposed Continental version of Burma, known as Korea. The wax model pictured here shows the car's sporting attitude and remarkably modern frontal appearance. (Courtesy Rolls-Royce Motor Cars Ltd.)

Built on chassis 61-B, this Korea prototype car had a Park Ward body. Note the sleek styling and Bentley wheeltrims. (Courtesy Rolls-Royce Motor Cars Ltd.)

stretched version to form the basis of the Austin 3-litre. Close examination of the 3-litre Austin shows it to share the same central body section as the Austin 1800.

As the BMC and Rolls-Royce collaboration developed there appeared a number of interesting designs, some of which were intended as aspiring sports cars. Some of the drawings and illustrations penned at the time show a distinct leaning towards Italian design style. Included in these is Alpha, a finely proportioned two-door sports coupé and the only existing quarter scale model of this project, built in fibre glass, now commands an elegant position in its stylist's office. Quaintly, the design of Alpha, which is over 30 years old, has more than a suggestion of current styling trends, especially with its tear-drop shaped headlamp lenses. Martin Bourne remembers with pleasure that the car was the winning entry in a competition for a design based upon BMC's 1800 chassis.

Martin is reluctant to take full credit for Alpha, however. The basic car was designed on a Healey chassis by two German design students, and in that form won a Farina-sponsored design competition. Farina built one and displayed it at various shows; when BMC turned the project down it was left to Martin to redesign the front end.

BMC's interest in Rolls-Royce

waned once it had negotiated the takeover of Jaguar and found for itself a niche in the luxury car market. The irony of the whole affair is that it mirrored in many ways exactly what W.A. Robotham had envisaged before the war.

Final developments and the SY project

As SY development continued from the remains of Burma and Tibet, so most of the road testing was carried out using the Bentley style of radiator, in order not to draw too much attention. While workshop tests were being conducted it was usual for a Rolls-Royce radiator to be substituted, especially when air-flows were being measured. By this time the new car's specification had been carefully defined: it would have Burma's 119.5 inch (3,035mm) wheelbase and the V8, 6.230-litre engine.

Five SY experimental cars were built before the Silver Shadow and T-Series Bentley went into production. A change was made to the project numbering system so that the experimental car's registration number corresponded with the chassis number. This allowed easier identification and was made possible by the introduction of a change in Britain's car registering system which, from 1963, employed a suffix letter following the index letters and numbers to denote the year of registration. As chassis numbers were suf-

fixed 'B' so an arrangement with the licensing authority was made to supply Rolls-Royce with a corresponding set of numbers. The first three SY experimental cars, which had chassis numbers 45-B, 46-B and 47-B, bore the corresponding registration numbers ALD 45B, 46B and 47B and were hand-built in the experimental department. Instead of the bodies being prepared by Pressed Steel or Park Ward, the panels were supplied by Airflow Streamline of Coventry. Successive test cars were built using bodies supplied by Pressed Steel and were therefore more representative of the production models.

Road testing 45-B revealed the now familiar problem of excessive vibration and noise from the suspension system. In an attempt to cure the fault the car was subjected to the torturous bump rig which simulated conditions with varying degrees of violence far in excess of conditions the vehicle was likely to encounter on the road. Tests were continued until engineers were satisfied with the results. 46-B was also subjected to the tortures of the bump rig and suffered substantial damage to the body.

The third SY experimental car, 47-B, was considered advanced enough in development to endure long-distance road-testing and was dispatched to France where road surfaces were considered especially poor. Test drivers in charge of the car managed to

A number of clay scale models were made in order to perfect styling. Here, ideas (codenamed Tibet and Burma 3) have been sculptured for both Rolls-Royce and Bentley versions of what eventually became the Silver Shadow and T-Series. (Courtesy John Blatchley)

attain remarkably high mileages and when the vehicle returned to Crewe it had covered over 21,000 miles (33,600km) without serious problem. Continuous testing did, of course, reveal minor faults and weaknesses but these were dealt with as they arose. There is evidence that road noise and vibration was still of some concern as noted when Dr Llewellyn-Smith and Harry Grylls tested the car.

The last two pre-production experimental cars were made available

as tooling for the new car was being arranged. 48-B, which was constructed with a Pressed Steel body, featured left-hand steering and was built to export specification. After a brief test period in France the car was taken to America and did not return until two weeks before launch of the Silver Shadow and T-Series to Bentley and Rolls-Royce agents at Crewe. A period of endurance testing in America was obviously considered of great importance, especially as this promised to be

one of the cars' strongest markets.

Despite attempts to keep identity of the Silver Shadow a secret, rumours of the car's existence began to abound as test cars were seen in the area around Crewe. The tests were under the direction of the chief test driver, Tony Martindale, who was well qualified for his job. As well as being a test pilot he had a formidable reputation for handling a motor car and could often be seen putting Silver Clouds and S-Series Bentleys through their

The SY project evolved from the Tibet and Burma projects, which were cancelled. The outcome was the Silver Shadow, seen here alongside a Silver Cloud III. 100 LG was a registration used for publicity cars. (Courtesy Martin Bourne)

Part of any car's development programme is crash testing and, as far as the Silver Shadow was concerned, this was under the direction of Jock Knight. This particular photograph was taken in 1968. (Courtesy Rolls-Royce Motor Cars Ltd.)

paces on narrow Cheshire lanes.

Crash testing was all part of the experimental programme and while this was costly it was nevertheless a very important part of the Silver Shadow's development. It was mostly possible to treat the exercise with some economy by using the same vehicle for crash purposes up to four times. Once a rear-end and side-impact crash had been staged, the car could be subjected to a front end collision before being dismantled. Trials were also carried out at MIRA, the motor industry's proving ground at Nuneaton, where tests were conducted on exhaust emissions, stability under tyre deflation and crash resistance. As road testing progressed, so pressure was put upon the experimental department to eliminate the problems that arose. One considerable difficulty had been to find a solution to the sometimes excessive road noise and vibration, problems far easier to cure when designing a car with a separate chassis. A monocoque shell, by virtue of its fundamental engineering, acts as something of a drum and what was needed was a suitable cushion between the bodyshell and the subframes. Other manufacturers would often use rubber blocks to perform this task but for a vehicle the size and weight of the Silver Shadow this was not good enough. Instead of rubber mountings Rolls-Royce used cylindrical pads of stainless steel wire mesh which, due to their density, acted like variable-rate springs. Developed by

Delaney Gallay, it was not the first time Rolls-Royce had used 'Vibrashock' mountings and had enough experience with them to insulate the exhaust system on the Silver Cloud III.

The Silver Shadow's exhaust system proved problematic in its installation and a solution to how to fit all four boxes that were required was difficult to find. Harry Grylls was eventually consulted and Jock Knight spent several hours with him in deciding which way the installation should be tackled.

Ride comfort was all-important and many hours went into perfecting the self-levelling suspension which provided unique stability. Not to be confused with the hydraulic suspension used by Citroën, Rolls-Royce utilised a completely conventional system of coil springs with double wishbones and anti-roll bar at the front and semi-trailing arms at the rear. Unlike Citroën's hydropneumatics which allowed the car to settle an inch or so from the ground when idle, the Rolls-Royce rested quite normally upon its coil springs. Neither was the self-levelling system inter-connected between front and rear. When driven, the car was kept at an even keel by hydraulic rams which operated very slowly; when the car was stationary the system worked ten times as quickly, instantly adjusting for changes in weight distribution as passengers got in and out, the petrol tank was filled or luggage loaded.

Interior styling was as important

as it had to reflect the sumptuousness of the luxury carriage which was all part of the Rolls-Royce and Bentley hallmark and reputation. When it came to the coachwork, modern techniques may have been accepted but the cosseting of passengers was a different matter. The same attention to minor detail was carefully applied and Martin Bourne can remember spending many hours ensuring perfection. Harry Grylls was anxious to test every stage of development and if the engineering director found the seats comfortable with his 6ft.3in. (1.95m) height, this was acceptable.

The specification of the Shadow was enticing: electric windows and seat adjustment, finest leather and veneer combined with the softest cloth. Driver satisfaction could not be ignored and close attention was paid to the attractiveness of the instrumentation and facia. Added to this exquisite comfort was the pleasure of driving a car with power steering and an effortless gearbox. Safety was always at the forefront: stopping power was provided by all-round disc braking, the system tested and proved until it surpassed even Rolls-Royce's own standard. An unfortunate design of the handbrake assembly, which protruded in a somewhat ungainly manner from under the dashboard, went for a long time unnoticed. When Bill Allen was allowed to use a test car for a weekend, a relation's wife climbed into the driving seat only to find the position of the parking

The Silver Shadow and T-Series Bentley were unveiled to Rolls-Royce dealers on 30th September and 1st October 1965. With self-levelling suspension, monocoque body construction and all-round disc brakes, these were the most innovative Rolls-Royces in over 50 years. (Courtesy Martin Bourne)

Below: At the Bentley T dealership launch, Dr Lewellyn-Smith (left of picture) explains some of its innovative features to his wife. (Courtesy Martin Bourne)

brake highly inconvenient for feminine attire. This was not an isolated complaint about the car's handbrake, and it is best left to Martin Bourne to illustrate the problem the styling and design departments were faced with:

"We did not want a great long lever with a noisy ratchet between the front seats, (not R-R!) or even outboard as in the days of yore. The only alternative was an exquisite variation on the Ford Popular umbrella handle theme, a beautifully styled and engineered pull-out job under the facia - just above the outboard knee. It worked beautifully but due to the weight of the car and the leverage required, handle travel had to be quite long. It was a relief when the foot-operated parking brake came along!"

By the summer of 1965 Rolls-Royce was ready to put the Silver Shadow and T-Series Bentley into production at Crewe. Changes were made at the factory to accept the new manufacturing process, which included inspection and rectification, where necessary, of bodyshells received from Pressed Steel. Upheaval at Crewe during this period was understandable and production was severely affected by the tooling process. Final testing of the prototype cars was carried out even while arrangements were being made to equip the production areas. Although it would have been legally inadvisable, the M6 motorway was used to conduct speed trials, during which, on occasion, it had been possible to reach over 120mph (192kph).

The Silver Shadow and the T-Series Bentley were scheduled for presentation to Rolls-Royce agents at Crewe at the end of September. Due to the number of invitations it was necessary to extend the event to two days and it was held on Thursday 30th September and Friday 1st October. During the two days 120 representatives of the home sales organisation were able to visit the factory and try for themselves the two demonstration cars made available for the occasion. At least three cars were on show but it was a Regal Red Silver Shadow, carrying the registration number 100 LG, and a T-Series Bentley, 1900 TU, which were used for testing purposes. These registration

There were smiles all round when the Silver Shadow was presented to Rolls-Royce dealers at Crewe. The identity of the gentleman standing alongside the car is unknown but Dave Tod (Scottish sales representative) can be seen standing beneath the porch and, in front of him, is Ian Vetch. (Courtesy Martin Bourne)

All smiles for the T-Series Bentley, too. Standing by the driver's door is David Buckle. (Courtesy Martin Bourne)

numbers, incidentally, were retained by Rolls-Royce for use on publicity cars. Reaction to the cars was undoubtedly mixed but there was no doubt that, in the agents' opinions, the models lived up to everything expected from a Rolls-Royce or Bentley.

Although the new car had now been launched there was no time to be wasted as constant improvement was an essential facet of manufacturing. It was business as usual in the design department as plans were underway to announce a two-door version of the Silver Shadow and T-Series. In addition, Mulliner, Park Ward was also preparing a spectacular drophead version.

ROLLS-ROYCE & BENTLEY

II

WITHOUT A SHADOW OF DOUBT

Relieved at the Silver Shadow's successful launch to its agents, Rolls-Royce got on with the task of building cars for sale. The design and test departments, instead of being able to relax their work schedule, were immediately subjected to further pressures. Solutions had to be found to the many problems which were consistently being encountered by the engineering department, service section and, ultimately, the customer. Testing the Silver Shadow and Bentley was, therefore, all the more important once the car was in production and, to this end, experimental cars were put through their paces with an even greater sense of purpose and urgency.

Derek Coulson (*DC*) and J. (Mac) Macraith-Fisher (*McF*) were responsible for development and can recall the many hours spent trying to find answers to particular problems, some of which, at the time, seemed almost insurmountable. Their first impression of the initial prototype car left them with a feeling of dismay: in their own words it was "almost appalling by Rolls-Royce standards". The striking feature about the original Tibet car had been its sheer size but when Burma came along it was an altogether better car, apart from being grossly underpowered. Performance was improved beyond all measure when the V8 engine was substituted for the 6-cylinder unit under the bonnet.

Road-testing the cars had produced some nerve-wracking experiences, especially when it came to stopping. A favourite venue of the test team for brake tests was Hopton Hill, near Belper in Derbyshire, where the gradient was of such steepness that if the car reached the bottom without incident it could be reasonably assumed the brakes were without fault. On a particular occasion Dr Llewellyn-Smith (Doc Smith), who was managing director at the time, experienced complete brake failure while trying out one of the experimental cars for a few days on extended trial. Mac Fisher remembers the incident well, especially following the telephone call from a very distressed Mrs Llewellyn Smith who had been in the car with her husband!

There is no denying that the de-

The last of the SY experimental cars, 50-B, was built in October 1966 and used for testing purposes. (Courtesy Rolls-Royce Motor Cars Ltd.)

sign and development departments at Rolls-Royce did experience some quite extraordinary difficulties which often took a long time to resolve. It might have been assumed by some that a car with the reputation of being "The Best In The World" would be devoid of teething troubles but, logically, it is the thorough way in which the cars were evaluated which earned them this epitaph. Derek Coulson had the task of taking experimental car 48B to the American continent where he subjected it to a gruelling two months' endurance ordeal. Wherever the car went, from Mexico to Arizona, it never failed to attract attention. Of course the car was disguised, the Rolls-Royce radiator having been replaced with something more like the Bentley's, and anything to suggest the car's origin was removed. Even so, the vehicle's identity was guessed by some, while others were sure it was the latest Jaguar. From Arizona the car was driven north to Canada and in total covered some 10,000 miles (16,000km) before it returned to Britain. On an another occasion a test car was being taken to France when it disgraced itself mid-Channel. When Derek Coulson, who again had the job of conducting the test, went to the car which he had left with the other cars on the ferry, he found that all the hydraulic fluid had leaked from the suspension system onto the deck.

Everything on a Rolls-Royce and Bentley - and the Silver Shadow and T-Series were no exception - was designed to be reliable. Should failure occur, it would usually happen at a slow rate, which would normally give the driver time to take compensating action or rectify the situation. An indication of Rolls-Royce engineering thoroughness was that every 100th engine was taken off the test bench, dismantled and checked for wear. Those engines were never reassembled for use in a production car.

The meticulous way in which the experimental cars were tested is evident from the trials each component was made to endure. If it failed, the item had to be examined carefully to find out exactly why and how it failed; the remedy was to have the item redesigned to prevent future failure. Tyres were a point in question: Rolls-Royce took products from three manufacturers, Avon, Dunlop and Firestone, and tested them beyond any reasonable level for safety at establishments such as MIRA. In order for Rolls-Royce to accept the tyres, the manufacturers had to ensure even higher than usual levels of tolerance.

For all its diligence the development team was economical in its costs and overheads but never frugal when ensuring quality and safety. In February 1966, a period when the department was at its busiest, the year's budget of £485,000 was underspent by £50,000.

The successful presentation to agents and journalists of the Silver Shadow and T-Series was, in effect, the pinnacle of achievement for one man, Harry Grylls. It was his drive and determination that had ensured the project would succeed. Grylls foresaw that, should Rolls-Royce remain a mainstream producer of luxury cars, there would be no alternative but to change the company's manufacturing traditions. From the outset he had full belief in electing the change to unitary construction and, in so doing, fully understood that a long and difficult passage lay ahead. He recognised there would be many problems but, nevertheless, was confident in the knowledge that his engineers had the ability and resource to overcome them all.

The cars displayed at the agent and press launch may have seemed to untrained eyes to resemble those early Burma and Tibet prototypes, but were, in effect, very different. Certainly some similarity of outward style and shape remained, even though the only parts of the car that did not undergo any alteration were the doors. In retrospect both Mac Fisher and Derek Coulson were adamant about the cars' development: there was just one way to evaluate a car and that was by *sitting* in it and *driving* it. It had taken a little over two years to perfect the cars to the stage they could be put into production: Rolls-Royce's board of directors had given their approval to the definitive car on 9th July 1963.

First reactions

Just what did motoring journalists think of the Silver Shadow when it was presented to them at Crewe in 1965, a few days before the London Motor Show? The late John Bolster, a Rolls-Royce enthusiast if ever there was one, tried the Bentley T-Series after having driven to Crewe in his 1911 Silver Ghost. The difference between the two

Apart from its radiator and dedicated badging, the T-Series Bentley was virtually identical to its Rolls-Royce equivalent. (Courtesy Rolls-Royce Motors Ltd.)

cars is hardly comparable, apart, that is, from the remarkable build quality applied individually to each car. The Silver Ghost had been advertised in 1907 as 'the most graceful, the most attractive, the most silent, the most reliable, the most flexible, the most smooth running six-cylinder car yet produced'. How the motor car had progressed!

About the Silver Shadow, John Bolster was emphatic in his report for *Autosport*. 'A remarkably comfortable ride' he noted; further comments indicated that it 'felt astonishingly small being a car of reasonable size' but as he got more accustomed to it he clearly admired the car and liked its style. Summing up in a simple statement he claimed it to be 'a new car for modern conditions'. That, of course, was exactly what it was: the car-owning population was increasing rapidly - in 1959 there had been one car for every ten persons, by 1969 the figure had doubled to a car for every five. The rapid increase in popularity of the motor car did not, however, mean that Britain was doubling its output of cars, rather that there was a large increase in sales of foreign-produced vehicles imported to Britain.

'New look for The Best Car In The World, now with a lower, wider radiator shell and less flamboyant body contours' was *Autocar's* caption to one of Rolls-Royce's publicity photographs. A Rolls-Royce or Bentley is usually associated with grace and good taste but hardly flamboyancy, which has a suggestion of vulgarity about it. It is true the Rolls-Royce and Bentley driver no longer sat head and shoulders above most other motorists and, compared to the Silver Cloud, the Shadow did appear somewhat smaller. Yet there was no mistaking the sheer elegance and finely sculptured features that made the Silver Shadow and T-Series Bentley a car totally superior to any other.

The Australian journal *Modern Motor*, with its report by Harold Dvoretsky, sought to impress its readers by suggesting that something of a revolution at Rolls-Royce had made the company face up to the 'hard, cold economic and competitive facts of life' and go the way of most other popular manufacturers. Intended, perhaps, as a slight jibe at the luxury car maker, any affront was soon dispelled when the article concluded: 'when Rolls decide to change, they change in typical Rolls-Royce style - perfection personified'.

The Silver Shadow was so unlike anything else Rolls-Royce had ever produced, it was only natural that it made the impact it did at Earls Court in 1965. It was quite usual for manufacturers of some of the more popular makes to parade their cars in 'cutaway' style to show how every part of the car fitted together, but never Rolls-Royce - until the Silver Shadow, that is. Prominently displayed, but well protected from those eager to try out for size a Rolls-Royce, a Silver Shadow with its doors removed proved just how accessible the new model was. It was not only a question of accessibility, of course, but more about comfort and sophistication, the sumptuous luxury of burr walnut and finest hide, deep-pile Wilton carpets and superlatively soft West of England cloth. Unfortunately, displaying a car in this manner had the effect of spoiling the smooth styling line so, instead, emphasis was placed upon door hinge brackets and lock mechanisms which were normally out of sight. This may have seemed a little distasteful to some devotees of the marque who considered it quite unnecessary to denude the car of its fine lines (bare chassis of Rolls-Royce and Bentley cars had, after all, once been a feature at some motor shows) but, in defence, the display was impressive. What is more, it attracted unprecedented attention and interest. (Around 1963/4, Rolls-Royce put a Silver Cloud III chassis - which is now at the R-REC Headquarters at Paulerspury - on the stand, which had never been done before. 'Very nice' said one journalist, 'but rather like seeing royalty in the nude...')

At £6556 the Silver Shadow cost almost as much as four Rover 3-litre coupés and almost two of Daimler's impressive Limousines. Those motor show visitors who were happily considering the purchase of a new car such as the Ford Anglia or Austin or Morris 1100, might have been wryly amused by the thought that they could have bought a whole fleet of ten of these cars, and still had a pocket-full of change, for the price of a single Silver Shadow!

Displaying the Silver Shadow without its doors was Crewe's way of showing how the marque had adapted to the motor industry's advancing technology. This Rolls-Royce was no longer a carriage with its separate chassis, the motor car for aristocracy, but a car for those customers that, quite simply, wanted the best that money could buy. It was a town car, a vehicle to be enjoyed, to be *sat in* and be surrounded by the monocoque bodyshell, itself the result of mass-production methods. The finery was also there to relish, of course: fold-down picnic tables, centre arm rests, electric windows and even powered front seat movement. Above all, the car had that distinct aura of opulence. In many ways the Silver Shadow was responsible for taking

some of the mystique out of Rolls-Royce ownership; in the eyes of some the car had been brought down to earth which, in a way, was only to be expected as Rolls-Royce was in the market to sell cars profitably.

There was no mystery at all about the fact that, whilst the Silver Shadow was being presented around Europe at all the major shows, Rolls-Royce did not have any examples of it to sell. This was particularly unfortunate as the car had been so well received that a healthy demand existed. Understandably, the company's agents and sales representatives were besieged by customers and dealers clamouring to get their hands on the new car. The board of directors at Rolls-Royce had been anxious to launch the Silver Shadow at the 1965 Paris Motor Show and not wait until the car was safely in production. The latter would certainly have meant postponing until 1966 any announcement of the new model and by then the first occasion the cars could have been seen publicly in Britain would have been the London Motor Show. There were a number of reasons for the delay in actually having cars to distribute to agents: firstly Rolls-Royce had to be sure the cars were ready for production and therefore as many teething problems as possible had to have been rectified; secondly, Pressed Steel had to be able to supply the bodyshells to the strict requirements of Crewe's engineering division.

The arrangement was that Rolls-Royce would supply Pressed Steel with

a hardwood model of the Silver Shadow/T-Series to formulate blueprints. Once this process was completed the design had then to be 'detailed' in order that the bodyshell could be built as an entire unit at Cowley. Admirers of the Rolls-Royce might have experienced grave concern at knowing the Silver Shadow's bodyshell was produced alongside, and by virtually the same process as, many of Britain's most popular cars. They need not have worried, though, as Pressed Steel not only produced the bodies for Morris and Austin and other cars in BMC such as Riley and MG, but it was also a major supplier to the Rootes Group, responsible for names as famous as Humber. As a company, Pressed Steel was quite accustomed to the demands made by Rolls-Royce and had been the major supplier of body components since the immediate postwar period, when the Mark VI Bentley, and later the Silver Dawn, were introduced, through to the late fifties and the Silver Cloud III. Moreover, quintessentially British car makers Jaguar and Rover body production was also down to Pressed Steel.

The cars made available to agents and motoring journalists for the models' launch at the end of September and beginning of October 1965 had been quickly prepared once the events were over and shipped to France in time to make their appearance at the Paris Motor Show. A couple of weeks later they had arrived back in Britain where they were made ready for dis-

This artwork of the Silver Shadow depicts the car's clean styling lines. John Blatchley, by utilising the monocoque shell to its full extent, managed to produce a car with as much interior space as the Silver Cloud but of a more compact design. (Courtesy Rolls-Royce Enthusiasts' Club)

play at Earls Court for the London Motor Show.

David Tod, Rolls-Royce's sales representative for Scotland, remembers potential customers offering amounts well over and above the car's selling price in order to get to the top of the waiting list. Any tactic was to no avail, of course, as Rolls-Royce certainly did not operate in such a manner. As sales representative, David was allocated a demonstration vehicle and had the privilege of receiving the first production T-Series Bentley, which was built on chassis number SBH1001. Those customers lucky enough to get hold of one of the early cars were in the fortunate position of being able to re-sell it, should they so wish, at considerable profit.

In recalling the initial affect the Silver Shadow and T-Series Bentley had on his customers, David Tod remembers some were a little apprehensive about the car. More used to being chauffeur-driven in the expansive style of the Silver Cloud and Silver Wraith, the Silver Shadow appeared, by contrast, rather confined. Some even found it impersonal. The Silver Shadow and T-Series Bentley were, of course, designed for a different generation; owner drivers who did not employ chauffeurs. In recalling those early days with his Bentley, David Tod remem-

bers prospective customers' reactions when he demonstrated the car to them. Any suggestion that it, or the Silver Shadow, was a lesser vehicle than previous cars was quickly dismissed. The car handled surprisingly well, much better than most thought it would, as long as, David recollects, performance was kept within the car's limits. There is no denying the handling had shortcomings, as was recognised by John Hollings (*Hgs*) who, as chief engineer of the company after 1968, revealed Rolls-Royce's engineering practice during the 1970s at the 1982 Sir Henry Royce memorial Foundation Spring Lecture.

With the Silver Shadow, Rolls-Royce was confident it could build upon the successes previously enjoyed in America with the Silver Cloud. Actively courting the American customer, production at Crewe was so geared that the first 249 cars were allocated to the home market and the 250th car was destined for the USA, leaving Liverpool on 1st March 1966. To please the American customers Rolls-Royce had to tread a careful path and, whilst it was important the car retained its very unique 'British feel', it also had to have a ride commensurate with that of the Cadillac. In contrast to European motorists, who generally favoured cars with firmer suspension, American drivers preferred the ride to be wallowing

and relaxed. The Silver Shadow's hydraulic self-levelling suspension was, therefore, seen as a very important feature. After the Silver Cloud there was the danger the Silver Shadow might have seemed rather small when compared to some of the huge American machines with their vast bonnets (hoods) and massive boots (trunks). In the event, there was no cause for concern as America welcomed the car with open arms. Rolls-Royce, incidentally, had taken the precaution of obtaining a Pontiac which was sent to Crewe to allow the development team to examine just what the American customers liked. Some commentators have since suggested that the Silver Shadow was a car expressly designed for the American market, built for America and sold to America.

The first detailed information about the Silver Shadow appeared in America early in 1966, courtesy of *Road & Track* magazine where, immediately and still some four years later, comparisons with Buick and Lincoln were made, not to mention Mercedes. The Rolls-Royce did have a 50,000 mile warranty, it stopped quicker than the American machines and was more manoeuvrable in traffic. As to the bottom line, some jealousy is evident and the following comment made by the journal almost certainly would not have pleased the directors at Crewe: '... a small manufacturer, is hard-pressed these days to match the standards set forth by the giant automakers - American or otherwise - and that Rolls-Royce can no longer be considered the completely magical motoring experience it had the reputation for in the past.'

America has always been an important market for Rolls-Royce, not only cars but aero engines too. During the Second World War Rolls-Royce Merlin engines were produced in America by Packard. Rolls-Royce, Inc. was established in Detroit where it manufactured the 'Rotol' blade rotator, a major component of a propeller hub assembly. In postwar years the car business moved away from being a distributor-based operation to one where Rolls-Royce, through Rolls-Royce Inc., became the direct distributor, so retaining for itself the distributor's profit margin. Direct distribution was made effective from 1st October 1964 and in charge of the operation was John Simonson and George Lewis. Simonson was appointed wholesale manager, the position he had held with the major Rolls-Royce distributor, J.S. Inskip Inc., of New York. It was while he was employed by Inskip that he had been particularly successful in building up a Rolls-Royce dealer network and therefore gained invaluable experience. Lewis was hired as sales manager, the position he had held with Ford of America as well as being one of that company's executives.

As in Britain, arrival of the Silver Shadow in America was watched with some concern: the Silver Cloud had been immensely popular and its replacement at first appeared very much less imposing. Marketing of the Silver Shadow took on a different theme: it was aimed at a much younger owner-driver. To a large degree the new marketing ploy was successful and demand quickly exceeded supply.

More than just a new car from Rolls-Royce, the Silver Shadow was part of the era popularly termed the 'Swinging Sixties'; the Beatles and the Mersey Sound, in vogue was flower power and Britain enjoyed a love affair with the Mini while the *mini skirt* symbolised the generation gap. There was social change, too: a Labour government had been elected under Harold Wilson after years of Conservative rule; fashion looked to Carnaby Street and John Lennon shocked society with his psychedelic Phantom V. James Bond became an international hero and drove an Aston Martin, while Boeing 707 airliners flew millions of jet-set holiday makers around the world to previously unknown resorts. On the motor racing circuit the world cheered on Graham Hill and Jackie Stewart; John Surtees was champion on two wheels as well as four and Britain mourned the death of Jim Clark in 1968 when his Formula Two Lotus crashed at the Hockenheim circuit in Germany.

The monocoque

In fundamental engineering terms the monocoque structure of a car is all about frequency and torsional stiffness and motor engineers had to achieve a frequency equal to that of the wheel rotation of the car when at full speed. In the case of the Bentley T and the Silver Shadow, this represented something above 1200 cycles, the speed of rotation of the road wheels at maximum revs. Suffice to say, the design team achieved over and above what was required.

The bodyshells were delivered to Pym's Lane from Pressed Steel by road transport where at once they were subjected to what was known by the Rolls-Royce's engineering department as 'white lining'. The bare unpainted shell was referred to as 'Body In White' and the department through which it passed prior to painting was known as the White Line, which was under the control of its foremen Fred Molyneux and Stan Burrell. Having had its protective wax coating - which had been applied before leaving the Pressed Steel plant at Cowley - removed, each bodyshell was given a rigorous inspection to check against imperfections, which were carefully noted. Correcting blemishes, the lead filling of any inconsistencies in the bodywork and repairs to any accidental damage sustained in transit were then carried out. All panel work was painstakingly examined and meticulously measured while every aperture was carefully checked for accuracy.

Each bodyshell received from Pressed Steel was made to the same precise specification. The adaption to either left- or right-hand steering was carried out at Crewe where, at the same time, all necessary drilling of

44

holes in the panels to accommodate instruments, wiring and mechanical equipment was completed. On average it took a total of two days to perform these initial tasks.

Where imperfections were found in the bodyshells rectification was always carried out in-house at Pym's Lane. The extensive reorganisation at Crewe necessary to facilitate production of the Silver Shadow, had allowed for installation of a completely new manufacturing process. Included was an up-to-date preparation plant which enabled complete bodyshells and panels to be finished to the highest standards. Alterations to the Crewe factory were undertaken throughout the summer of 1965, the pinnacle to events which had been planned for what, to Macraith Fisher's mind, had seemed an interminably long period.

At first glance the bodyshell of the Bentley T seemed almost indistinguishable from that of the Silver Shadow, a fact which resulted in a dramatic downturn in Bentley sales as most custom-ers opted for the prestige of a Rolls-Royce. On closer examination, however, it was evident that the bonnet panels had a slightly different pressing which allowed for each marque's shape and style of radiator. Although the main frame of the monocoque was

constructed from steel, the doors, bonnet and boot lid were formed from aluminium, representing a substantial weight saving.

The build process of the Silver Shadow and T-Series cars included protection against corrosion. Water traps, as far as possible, were eliminated and vulnerable areas underneath the cars either galvanised or stove enamelled. Something like 400 parts were so treated, on top of which approximately 70lbs (32kg) of under sealant was applied in two stages. Stainless steel was used for many of the components, replacing chromium plat-

To accommodate Silver Shadow production extensive reorganisation was necessary at Crewe and included the establishment of a body preparation plant. Rubbing down the bodywork was only a part of the process which culminated in 15 coats of paint being applied. (Courtesy Rolls-Royce Motor Cars Ltd.)

ing where possible.

A new paint shop which facilitated total immersion of the bodyshell to apply the primer and protect against corrosion was all part of the new manufacturing plant. Fifteen coats of paint were applied to each bodyshell and the whole process included numerous inspections, washing down and degreasing. After each session in the drying ovens the surface of each car was meticulously examined for any blemishes which were marked by pieces of white tape. Only when the finish was perfect could the final painting operation be carried out - the steady and painstaking application by hand of the coach lines.

The average time it took to build a car was 12 weeks but, depending on whether the customer required any special features, could take longer. The whole process for each vehicle was fastidiously recorded in a log file containing anything up to 40 pages. Details of each car are, incidentally, kept as a permanent record and these are currently available at the Rolls-Royce Enthusiasts' Club headquarters at The Hunt House at Paulerspury in Northamptonshire. As the car passed through each section any defects were noted and put right before the car was allowed to continue. As the vehicle had to be signed off at each stage, any flaws or imperfections could be immediately traced to a department or individual. Quality control was the domain of Frank Dodd while Doug Fox was in charge of testing. Doug's son, John, incidentally, spent a number of years as manager of the experimental department.

The assembly hall was divided into three areas: pre-mount, mount and post-mount. As the bodyshell and subframe assemblies (built on their own assembly lines) progressed along the route upon specially designed trolleys, so the components were fitted at pre-

The main car assembly shop. In the foreground engines await installation. (Courtesy Rolls-Royce Motors Ltd.)

Lowering a V8 engine onto a subframe assembly. The photograph was taken in 1967. (Courtesy Rolls-Royce Motors Ltd.)

kept on the vehicle and the opportunity to make any adjustments necessary. Only when the test drivers were sure the cars performed as they should, and were without any squeaks or rattles, were they dispatched to be valeted, cleaned and waxed.

Most evident about the 'platform' of the Silver Shadow were the car's two massive subframes, each appearing more like a chassis structure. The front sub-assembly consisted of forward and rear pick-up points, which supported the main structure and were connected just ahead of the A-post rear of the wheelarch, immediately ahead of the front wheel assemblies. The rear subframe was similarly constructed and connected to the body on each side of the car both ahead and behind the D-post.

Supported upon the front subframe was the engine and gearbox, steering assembly, front suspension and braking system. At the rear of the car, the sub-frame supported the rear axle and final drive, suspension and brakes. Forming a means of insulation between the sub-assemblies and the bodyshell, *Vibrashock* mountings, (universally known as 'pan scrubbers' as

arranged stations. It was at the mounting stage that the bodyshell and front and rear sub-frame assemblies met with each other and were made to resemble a motor car. The jig that was used to knit the car together was known affectionately as the 'Queen Mary', no doubt due to its impressive structure. It was at the post-mount stage that

such items as wheels, tyres and exhausts were fitted to the car before being sent to the finishing shop for fitting out.

All cars were road-tested minus the radiator shells, bumpers and trim items such as hub caps, for as many as 150 miles (240km) but usually much less, which allowed a close watch to be

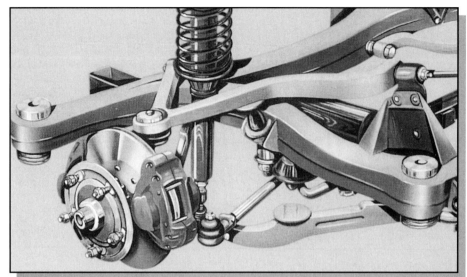

Left: The front sub-frame assembly. (Author's collection)

Below: Rear subframe assembly. (Author's collection)

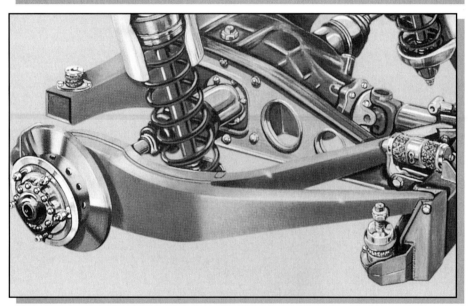

but it is here any fundamental likeness ended. Whereas the system employed on the DS was totally hydraulic - no other form of springing was used - and height settings were selected by the driver, that used by Rolls-Royce and Bentley merely kept the car at a specific level and was supplementary to a conventional system.

Hydraulic system

To explain the Silver Shadow's suspension it is necessary firstly to examine the car's hydraulic system which controlled three distinct functions: pressure generation, braking and height control. Each function had its own circuit which, in turn, was made up of a number of components.

Taking each function separately, the pressure generation circuit comprised a fluid reservoir, brake pump and two accumulators, the latter incorporating a valve and sphere assembly. The reservoir, which enabled a head of hydraulic fluid to be maintained, as well as allowing the fluid to expand as it heated, was, in fact, two units, front (number one system) and rear (number two), supported by a single housing. The fluid passed to the brake pump where it was contained between the outer casing and the body of the pump. The pump unit, which was bolted to the engine tappet cover, took its drive from a pushrod operated from the camshaft and displaced the fluid, so forcing open a non-return valve. The fluid was then passed to the accumulator which stored and regulated the pressure. After use the fluid returned to the reservoir. Originally

they were made of stainless steel wire) which had been developed by Delaney Gallay prevented to a large degree road noise and vibration from being transmitted to the car's interior via the suspension system. Incorporated as part of the car's suspension, the self-levelling device compensated for changes in vehicle load and also allowed a certain freedom of fore and aft movement. Whilst providing a stable ride, the horizontal compliance also induced a degree of vagueness about the car's steering which provoked a considerable amount of controversy and criticism. Ultimately, of course,

this was rectified.

The hydraulic system and, ultimately, suspension layout, was considered by motoring journals at the time as probably the most innovative of all the features of the car. Motoring correspondents paralleled it, rather erroneously, with that of the hydropneumatic self-levelling system adopted by Citroën for the large D-series cars, but closer examination reveals distinct differences. Certainly there were similarities in as much that Rolls-Royce used high pressure accumulators and pressure regulating valves built under licence from Citroën,

Work in progress in car assembly, Crewe. In this busy scene, cars are having trim items fitted. (Courtesy Rolls-Royce Motor Cars Ltd.)

Awaiting body furniture, the car in the foreground has left-hand steering. Most of the cars in the assembly shop seem intended for export. (Courtesy Rolls-Royce Motors Ltd.)

sphere fill with fluid. Pressure was allowed to reach 2500 psi before it returned to the reservoir.

The braking system of the Silver Shadow was a sophisticated affair which incorporated all-round disc brakes (Rolls-Royce, incidentally, was the last of the major car manufactur-

both accumulators were located at the rear of the engine on the left- hand side of the car, but after 1975 the forward accumulator was moved to the front right-hand side of the engine. The accumulator comprised a sphere charged with nitrogen at 1000 psi and incorporated a diaphragm which allowed fluid to be contained in the upper segment. As the pressure of the fluid rose and equalled the pressure of the nitrogen, so the diaphragm dropped, letting the

Getting to grips with the job: adjustments are being made while a car is on the overhead ramp. (Courtesy Rolls-Royce Motors Ltd.)

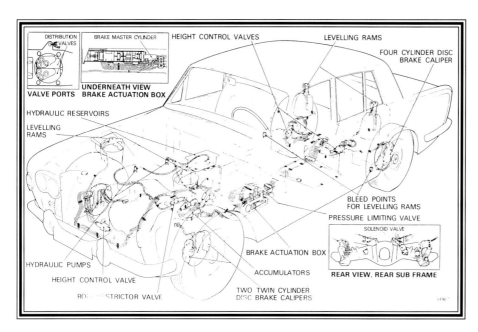

The Silver Shadow's hydraulic system was necessarily complex, as depicted here. (Author's collection)

ers to adopt disc braking). Not only had drum brakes proved extraordinarily efficient but the company needed to be sure the system it had decided upon was without problems and did not suffer the inherent brake squeal that had so often plagued other car makers. As can be expected a lot of time and effort was taken by the development department which had worked alongside component suppliers to establish a suitable design. The front brakes comprised two calipers on each wheel and the rear brakes a single caliper on each wheel, all brakes utilising two sets of pads, those on the rear wheels being larger than those on the front. The braking system, which was activated by two distribution valves and operated by pressure on the brake pedal, supplied hydraulic fluid under pressure at 2500 psi via two independent circuits, each with identical components. In addition a pressure limiting valve was fitted to the rear brakes and prevented the rear wheels from skidding under emergency stop conditions. As if this were not enough, a third circuit was provided which took the form of a conventional system and utilised a master cylinder in the usual manner which operated on the rear wheels only. Had the conventional circuit not been added the braking sys-

tem would have had very little 'feel' to it and the driver would have been subjected to the 'full-on' effect so often experienced with Citroën's hydropneumatic system.

The ratio at which the braking system operated was 47 per cent from the forward pump, which divided the effort between front and rear at a ratio of 31 to 16 respectively, a further 31 per cent of the total for the front only, and 22 per cent in respect of the conventional system. The handbrake (in America the parking brake), which provided, in effect, a fourth braking circuit, worked in a quite conventional manner, operating mechanically on the rear brakes which were also served by the master cylinder. For the American market a foot-operated parking brake

The hydraulic system of the Shadow and T-Series cars could reasonably be described as a plumber's nightmare! (Author's collection)

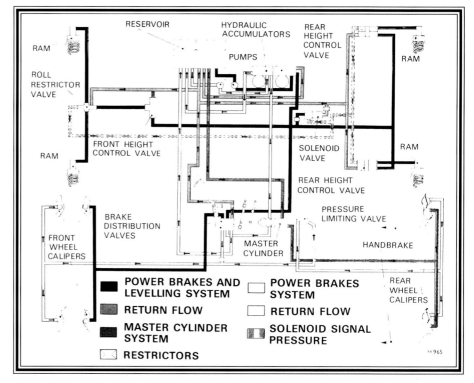

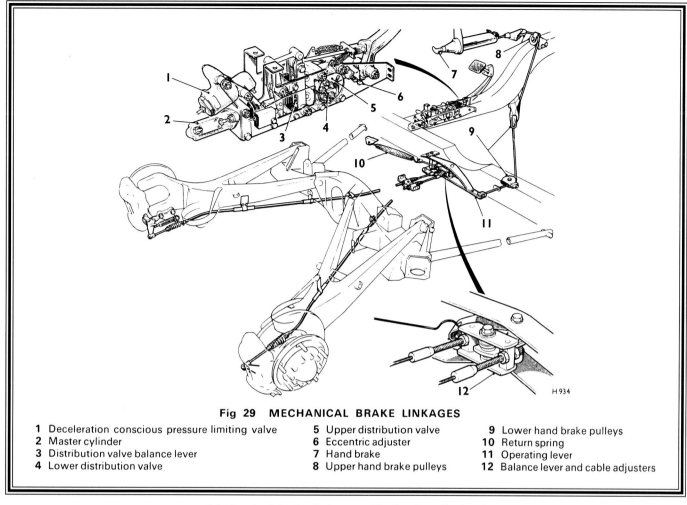

Fig 29 MECHANICAL BRAKE LINKAGES

1 Deceleration conscious pressure limiting valve
2 Master cylinder
3 Distribution valve balance lever
4 Lower distribution valve
5 Upper distribution valve
6 Eccentric adjuster
7 Hand brake
8 Upper hand brake pulleys
9 Lower hand brake pulleys
10 Return spring
11 Operating lever
12 Balance lever and cable adjusters

Mechanical brake linkages. (Author's collection)

eventually replaced the more usual umbrella type handle specified for British and European production and was positioned under the facia ahead of the driver. In Britain anything other than a hand-operated lever to apply the parking brake caused a great amount of confusion: the hydropneumatic DS Citroëns similarly employed a foot-operated system which, whilst appreciated by the French, seemed to provoke endless criticism elsewhere.

Rolls-Royce specified Girling calipers with 11 inch (280mm) discs which incorporated in their peripheries a band of stainless steel wire. This was designed to dampen any brake squeal, a factor which had troubled the development team for a great deal

of time and which had prevented earlier adoption of disc brakes. A safety feature later built into the system was the provision of facia-mounted warning lamps which indicated either a loss of hydraulic pressure or that the brake pads were worn to the extent that they required changing.

The height control system consisted essentially of four major components: height control solenoid, solenoid restrictors, height control valves and levelling rams. The function of the system was to keep the car at a constant level whatever load was carried and however distributed, and had nothing to do with wheel movement or road surface. Had the Silver Shadow - a particularly heavy car at 40.6 hun-

dredweight (4546lbs / 2062kg) - been designed to have conventional springing which still gave the soft ride required, there would have been the need to compensate for those occasions when extra loads were carried. Had, however, the conventional springing been able to carry all the weight asked of it, it would by necessity have been very firm. The quality of the ride under these circumstances would have suffered, unless the load had been evenly distributed. Since it was not feasible to satisfy both conditions the self-levelling device offered suitable automatic adjustment at all times, whether the car was on the move or stationary.

The height control valves operated at two speeds and, under all driv-

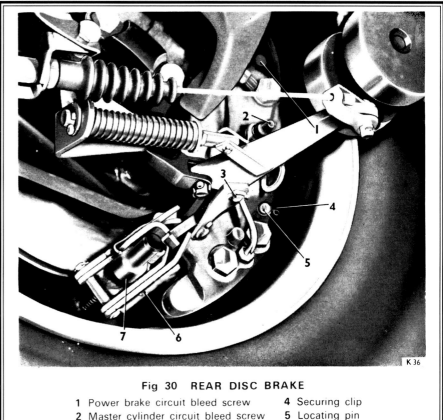

Fig 30   REAR DISC BRAKE

1 Power brake circuit bleed screw   4 Securing clip
2 Master cylinder circuit bleed screw   5 Locating pin
3 Disconnecting point   6 Hand brake pads
7 Adjusting ratchet seal

ing conditions, did no more than compensate as the fuel level reduced in the fuel tank. Once the car was at a standstill and weight distribution rapidly changed due to passengers getting in or out, or the fuel tank was refilled, the valves operated ten times faster. Three height control valves operated the four hydraulic rams at the front and rear of the car. A single valve, which was bolted to the front sub-frame just under the radiator, responded to rotary movements of an anti-roll bar. The two were connected by a linkage which adjusted the two forward levelling rams. Built into the hydraulic circuit was a roll restriction valve which, as well as dividing fluid between the two rams, restricted any crossflow between them. Without the restrictor, fluid would have drained to either side of the car causing it to roll alarmingly. Two separate valves, bolted on to the forward assembly of the rear sub-frame and linked to the trailing arms, operated the rams on the rear axle. It may have seemed simpler to install a height correction valve on each wheel but with this method it would have been virtually impossible to obtain the correct balance. In the event of rapid redistribution of weight, such as passengers getting in or out, the solenoid valve was activated to provide almost instant levelling, made possible by sensors located in the gearshift, when neutral was selected, and on the courtesy lights, which denoted a door opening.

The rams which operated the rear levelling device had a 3 inch (75mm) stroke, whilst those located on the front suspension had a movement of just 1 inch (25mm). This meant that the more weight the car carried the further the rams had to extend and, conversely, when the car was unladen travel was reduced. No more than a piston inside a cylinder with a seal, the rear rams were bolted on to the body of the car at the top of the coil springs and connected to the isolating cones which supported the rear shock absorbers and springs.

To ensure the self-levelling system worked efficiently and the valves were prevented from working quickly when they should have been operating slowly, solenoid restrictors constantly checked the hydraulic pressure. The levelling device, whilst functioning perfectly well at the rear of the car, did experience some early problems on the front axle and it was this which led to claims of steering vagueness. Compared to the rear levelling, the front

device had very little work to do and it was decided to dispense with it. An official modification kit was eventually supplied by the factory and with it there were full instructions on how to disconnect the front system. It is unlikely that many cars now exist which have not had this modification carried out.

Compared to the hydraulic self-levelling, the car's mechanical suspension layout appeared quite conventional: independent, with double wishbone geometry, at the front together with coil springs and Girling telescopic dampers; at the rear, independent suspension again but with single trailing arms, coil springs and Girling telescopic dampers. It is no wonder that the series of hydraulic pipes required to disperse the suspension fluid quickly (similar to the hydropneumatic Citroëns), became known as a plumb-

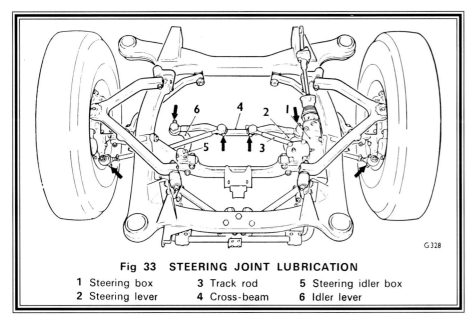

Fig 33 STEERING JOINT LUBRICATION

1 Steering box	3 Track rod	5 Steering idler box
2 Steering lever	4 Cross-beam	6 Idler lever

er's nightmare!

When introduced, the Bentley T and Rolls-Royce Silver Shadow models were specified with three levels of suspension. The American market demanded cars with a much softer suspension than would normally be required in Britain and Europe. As a result cars exported to the USA were provided with relatively soft springing more in the style of American produced vehicles. For the home market and Europe, the springing was made firmer, a benefit of which was markedly improved handling. The third suspension level was somewhat obscure and was reserved for those cars whose destinations were known to have the worst possible road surfaces. In such circumstances where it might have been possible to cause serious damage to any normal suspension, the springing was made considerably stiffer to withstand violent shocks. Cars with this 'colonial suspension' were usually destined for Australia and African countries.

Another feature new to the Silver Shadow and Bentley T was the steering mechanism. Instead of using its own powered steering, Rolls-Royce bought in the American Saginaw system which used a recirculating ball steering box instead of the Marles type of cam and roller specified for the Silver Cloud. As with the Silver Cloud, hydraulic pressure was obtained from a belt-driven Hobourn-Eaton pump but within a couple of years this was superseded by a Saginaw pump. The Saginaw steering had certain advantages as it utilised an internal ram, as opposed to the external type of earlier cars. As a whole, the Saginaw system was not only lighter than had been experienced with the Silver Cloud and S-series Bentley but, additionally, allowed a tighter turning circle of 4 turns lock-to-lock instead of 4.3. Even so, Harry Grylls' 'sneeze factor' was still evident - this was a safety margin in the event of the driver's attention being distracted (such as whilst sneezing), or in case of slight error when a twitch of the wheel might have caused other cars to wander off course. Ironically, soon after Harry's

Apprentices fitting the V8 engine to a sub-frame. The picture was taken in 1965. (Courtesy Rolls-Royce Motors Ltd.)

retirement the turning circle was reduced even further.

With safety as the main concern a collapsible steering column was designed, the first time such a device had been used on either a Rolls-Royce or Bentley. The column was kinked at its division which made it all the easier to install. For the first time also the steering wheel, which was 17 inches (432mm) in diameter, had just two spokes, instead of three as previously, although the rim retained the same dimensions as that fitted to the Silver Cloud. A central horn push was provided but *Autocar's* test driver was clearly not impressed with the two-spoke wheel, much preferring the previous three-spoke version.

Engine and transmission

Under the bonnet of the Silver Shadow and Bentley T was the familiar 6230cc V8 engine developed in 1959 which first saw service on the second series Silver Cloud and the Bentley S2. A V8 configuration had been decided upon quite early when choosing on a replacement for the faithful 4887cc straight six; the straight eight B80 engine had been a contender but was considered far too bulky. Apart from being fitted to the few Phantom IVs which had been produced, the B80's role had previously been to power military vehicles and it was used for industrial purposes. Again Harry Grylls looked across the Atlantic to Detroit to see what the Americans were doing: he saw and appreciated the lightweight V8s being fitted by Cadillac and Chrysler and acknowledged this to be

the route Rolls-Royce should take.

Before committing to the V8 layout, some consideration was given to alternative configurations. Rolls-Royce had previously developed a V12, which had been used in the Phantom III in the late thirties, and it was this principle that manufacturers such as Ferrari and Jaguar adopted. Daimler chose the V8 route for its Majestic Major and DR450 limousine, the 250 saloon (the Jaguar MKII lookalike) as well as the desirable SP250 Dart sports car. The V8 was specified for the big Fords also, and used by some specialist car makers such as Jensen who opted for the Chrysler unit.

It must be appreciated that when development of the Silver Shadow first began, it was the 6-cylinder engine that was considered. In adopting the Silver Cloud's V8 the opportunity was taken to implement a number of changes to the engine which suited the design of the new car. A new combustion chamber shape, made possible by redesigning the cylinder heads, proved more efficient and, as a result, improved maximum output by something like 2 per cent. Rolls-Royce, of course, never reveals the bhp rating of its engines, choosing instead to describe it merely as 'adequate'. Sparkplugs - which previously had been almost inaccessible beneath the exhaust manifolds - were made more readily available. In its previous form, engine accessibility was so dire that panels had to be provided in the inner wheelarches of the Silver Cloud just to facilitate sparkplug removal. Even considering the modifications carried out on the

engine, underbonnet space in the Silver Shadow was extremely tight. Maintenance and servicing, nevertheless, was made decidedly easier but there were few owners who concerned themselves about such detail ...

The increase in power was, to a great extent, necessary to drive the car's sophisticated hydraulic system as well as the plethora of ancillary components. Torque was also improved and a maximum speed of somewhere between 115mph (184km/h) and 118mph (189km/h) was possible. Incidentally, the best the Silver Cloud III could manage was 117mph (188.3km/h) which was impressive in view of the car's size.

Extensive use of aluminium alloy was made to form the crankcase, cylinder heads, timing covers and inlet manifolds. The 'wet' liner principle was used which meant that the cast iron cylinder liners were in direct contact with the coolant. Rolls-Royce had no concerns about the engine's specification: the valve gear was of the overhead in-line type with pushrods and rockers while hydraulic tappets completed the package. Two levels of compression ratio were offered: 9.0 to 1 for 100 octane fuel and, optionally, 8.0 to 1, for less refined fuels. Carburation was provided by two diaphragm type S.U. HD8 carburettors while two electric S.U. pumps fed the fuel supply. In order for the car to have a low bonnet line it had been necessary to modify the radiator header tank and, for the same reason, the air cleaner was relocated to inside the right-hand front underwing. Flexible trunking from the

air cleaner led to the carburettors.

To those who appreciate and understand Rolls-Royce's commitment to engineering, the V8 engine is a work of art. Weighing something like 30lbs (14kg) less than the 6-cylinder it replaced, it featured almost 30 per cent more in the way of swept volume. The design of the engine was conventional enough as a V8: it had a five bearing crankshaft and the camshaft was positioned neatly between the two cylinder heads, within the vee. Totally refined, as was to be expected, the unit promised a lifetime of smooth running as long as it was respected and carefully maintained.

In deciding the type of gearbox for the Silver Shadow, Rolls-Royce faced something of a dilemma. The automatic gearbox used in the Silver Cloud and S-Series Bentley had proved itself beyond all measure but, in the light of competition from new designs emanating from America, was showing its age. Although the gearbox, with a certain amount of modification, was acceptable to the British market, it had less chance of being appreciated in America. The costs in developing an all-new gearbox would have been prohibitive in terms of both time and expense, so it was decided that right-hand drive cars would have a refined version of the existing General Motors hydramatic box. Whilst of American design, the GM four-speed gearbox for Rolls-Royce and Bentley cars was actually assembled by Rolls-Royce at Crewe. For left-hand drive cars, and principally the American market, Rolls-Royce undertook to fit a completely

new gearbox to the Silver Shadow and Bentley T. In seeking a suitable unit Ivan Evernden did not have to look further than Detroit and the new GM400, again from General Motors. The gearbox, which was a very refined unit and totally smooth in operation, proved fully up-to-date in specification. Acknowledged as being just about the best available, it had already been chosen by Buick and Cadillac for their 1963 models. The first experience Rolls-Royce had of the GM400 box was when it had been installed in an experimental car as early as November 1964. It was fitted to 48B which at the time was undergoing trials. Torque was increased so much that, under tests requiring full-throttle starts, a halfshaft was snapped.

The new GM400 gearbox, known as the TurboHydramatic, had a number of modifications compared to the old four-speed type. It was fitted with a torque converter and three forward speeds, which did away with the previous fluid coupling, and second gear had just about the same ratio as the old unit's third. Performance was definitely better than that of the four-speed box, with first gear being a little quicker. Overall speed through the

gears was superior. Absence of the low-ratio bottom gear ensured an absolutely smooth operation which was mirrored throughout the gear range. Of interest is the fact that Rolls-Royce decided against building the GM400 at Crewe but instead bought in the gearbox as a complete entity and modified it to accept electronic actuation.

As for the faithful old four-speed gearbox specified for right-hand drive cars (it was intended that all Silver Shadow and Bentley T models would eventually receive the new GM400 box), some quite serious modifications were made. The main casing, instead of being formed from cast iron, was made from aluminium alloy, as were many of the internal castings; to add to its smoothness, a freewheel device was incorporated with first and second gear ratios but it remained possible, by selecting the hold on second gear and thereby making the freewheel inoperative, to obtain engine braking if required. A further refinement was the ability to change gear electrically, instead of having direct linkage, and a finger-light lever positioned on the steering column actuated a motor bolted on to the rear of the transmission casing. Mechanical links were, of course, provided to the selector levers in the usual manner but the refinement meant that physical gear selection was eliminated.

The reason for electric actuation extends back to the fifties when the Silver Shadow was conceived. In that period it was fashionable to allow front seat occupants to leave the car from either side, which was especially useful if the driver had parked against the nearside kerb and, not wanting to step out into the traffic flow, could simply slide across the front seat and leave by the passenger's door. As centrally mounted gearchange levers presented an obstruction, a popular solution was to place the gear selector on the steering column. A number of snags were associated with column change such as rattles and difficulty of operation, but, by placing a selector switch at the top of the column and connecting it to the actuator on the gearbox casing by nothing more complicated than an electrical cable, these problems could be eliminated.

Jock Knight, who was involved in developing a suitable electrically actuated gearchange for the Silver Shadow, was aware that great care had to be taken over the design and choice of materials for the electrical contacts of both switch and actuator. It was important that effective sealing of the actuator was achieved to protect against corrosion from water and road salt ingress; also that it was properly ventilated. Several mechanisms were mounted on experimental cars undergoing endurance testing and, in addition, switches and actuators were

tested on specially prepared rigs. One particular mechanism was installed in an environmental test chamber where it completed several million gearchange cycles while being exposed to extremes of temperature and humidity, including submersion in brine.

In the case of electrical failure it was still possible to get the car home. Rolls-Royce engineers had thought of everything as, included in the car's tool kit, was a specially-designed tommy bar which, when inserted into a hole in the top of the gearbox, allowed direct manual changes to be made. Needless to say the carpet over the gearbox housing had first to be removed. As a matter of interest it had been necessary to use the emergency tommy bar early on while testing experimental car 49B in France. The electrical gearchange had failed and the test team happily continued using the direct manual change until replacement parts were delivered from Crewe. After a few years' production with virtually no failures, the 'get you home' device was abandoned.

Instead of the two-piece propeller shaft which had been fitted to the Silver Shadow's predecessors, a one-piece affair was now specified. Not only did this add to the car's overall handling (the shaft connected with the differential which was fixed to the body shell, increasing the stiffness of the chassis) but also enabled a reduction in the vehicle's body length. The hypoid-bevel final drive unit was supported upon one of the two vertical crossmembers which basically made up the rear subframe; the other

crossmember supporting the suspension trailing arms. On the driveshafts could be found constant velocity universals of the ball and trunion Detroit type; on the outer shafts the usual Hooke-type Hardy-Spicers were specified.

Under the bonnet and out of sight (out of mind also for a great many owners!) was a plethora of ancillary equipment which included the cooling system comprising two thermostats and 28 pints (16 litres) of coolant, an oil filter - which had to be changed every 6000 miles (9600kms), and hydraulic pumps as well as all the associated belts, hoses and electrical wir-

All Silver Shadows and T-Series Bentleys had gearboxes with electric actuation. On early cars a 'get-you-home' device allowed manual selection of the gears in the event of electrical failure. (Author's collection)

Fig 8 MANUAL SELECTION OF GEAR RANGES

1 Tommy bar   2 Sealing cover

ing. The battery, a 12 volt 64 ampere model, was positioned in the boot on the nearside between the wheelarch and rear panel, while the alternator was a 35 amp Lucas unit. The engine sump held 14.5 pints of oil (8 litres) and the gearbox 24 pints (13.6 litres). The final drive held 4 pints (2.3 litres) of oil and the fuel tank had a capacity of 24 gallons (28.8 US gallons, 109 litres) of fuel. When just 3 gallons (3.60 US gallons, 13.64 litres) remained in the tank a warning light illuminated on the facia. Greasing of six grease points had to be attended to every 12,000 miles (19,200 kms).

Both inside and outside the Bentley and Rolls-Royce a most notable feature was the car's quietness. Excellent sound-proofing ensured occupants were spared mechanical and traffic noise and the finely engineered exhaust boasted no less than four silencers, and stainless steel at that.

An example of absolute sophistication was the electrical system which, on many more humble cars, would have seemed beyond all imagination. Not only were the window lifts electrically powered but also the seat adjustment. Should a headlamp fail, a secondary circuit automatically and instantly illuminated the stand-by filament and red lamps were installed in the rear doors to provide warning to following traffic. The headlamp safety circuit was introduced after Harry Grylls, who had been driving an experimental car, experienced a floor-mounted headlamp dipswitch explode, leaving its entrails lying on the floor and Harry with no headlamps, a fright-ening experience he did not want a customer to suffer. The solution was to provide the headlamps with an auxiliary circuit, in addition to the normal one, whereby in the event of an open circuit failure, a relay switched in a second circuit. In the event of a short circuit failure, a relay operated the headlamps intermittently, thus allowing the driver to bring the car to a halt with some semblance of lighting.

Today, it is taken for granted that a rear windscreen de-mister will be fitted as standard equipment in almost any car and electrically-operated aerials are commonplace, as is remote control opening of the fuel filler cap. In 1965 such features were luxuries, however, and certainly air-conditioning in a car was almost unheard of. The heating and ventilation system proved to be an elaborate affair and, of course, sundym tinted glass was a standard feature when air-conditioning was specified. The main heating and ventilation air intake was positioned on the scuttle, immediately in front of the windscreen and, in addition, two supplementary air intakes were placed at the front of the car, directly under each of the double headlamp units. The slim rectangular grilles positioned each side of the radiator provided cool air for the cabin which was fed to outlets on the side scuttle panels (generally known as 'ankle freezers') alongside the front occupants' legs. These were deleted one at a time during the early life of the car although the grilles remained at the front for some time. For the heating system, two four-speed blowers forced temperature-adjusted warm air into the car's interior, one for each side of the car, and the airflow could be directed either upwards, to demist the windscreen, or downwards, for cabin warmth. Heating ducts were also positioned in the rear compartment for the comfort of those passengers but, surprisingly, these could not be regulated or closed independently and were, in fact, controlled to provide the same output as the lower front heater outlets.

In its appraisal of the Silver Shadow, *Autocar*, in the autumn of 1965, was critical of the overall heating system in as much that it lacked any thermostatic device by which to hold the temperature. The report also considered that, had extractor vents been provided, these would have allowed a through-flow of air. It seemed a pity to have to open windows in an attempt to aid ventilation and heating efficiency. Much modified to that which had been specified for the Silver Cloud and S-Series Bentley, the air conditioning system was controlled by an evaporator unit and heater matrices positioned behind the facia. On the previous model these main components were housed in the valence of the front right-hand wing. The compressor was belt-driven from the crankshaft and the condenser located immediately ahead of the radiator.

Craftsmanship and excellence
Without doubt the single most recognisable hallmark of any Rolls-Royce is its radiator and, in this respect, the Silver Shadow was certainly no exception. To untrained eyes the

An air conditioning unit being assembled at Crewe. (Courtesy Rolls-Royce Motors Ltd.)

imposing radiator shell appeared quite flat and vertical when, in fact, it was neither. It took twelve highly skilled metal workers, all of whom had at least 25 years' experience in serving the marque (the company's own publicity material referred to them as acolytes) to make the hallowed radiators and Rolls-Royce rightly believed that no-one outside the factory could do the job as well. Looking more like silver or silver plate than the stainless steel it was, the perfect finish was the result of hours of polishing. The shell may have seemed composed of a single piece but, in fact, comprised no less than eleven separate items of steel, all of which were approximately 1/16th inch (2.5mm) in thickness, and worked until the edges were perfectly mitred.

Such precision may seem excessive: to other manufacturers perhaps, but not Rolls-Royce, where only the best was good enough. Traditional methods of metal working were employed, such as using age-old soldering irons heated over an open flame. The same applied to the *Spirit of Ecstasy*, possibly the best-known car

mascot in the world. In 1911 Charles Sykes ARBS created the emblem for Rolls-Royce as a commission and thereafter, for a number of years, cast each

figure - the base metal was an alloy of copper, either brass or bronze - himself. The model for the emblem was Eleanor Thornton, the secretary to the first Baron Montagu of Beaulieu and, surprisingly, this prestigious item was not standard equipment until after the Second World War. Before then it was supplied as an optional extra. Today, of course (and the same applies to the Silver Shadow), the mascot is crafted by Rolls-Royce's own engineers. Standing proudly on the radiator shell top tank, the figure conforms to current safety regulations and is spring-loaded so as not to cause injury. As part of the company's quality control measures

A Silver Shadow receiving final inspection. (Courtesy Rolls-Royce Motors Ltd.)

even the spring-loading mechanism of the mascot was checked by having a wooden ball, encased in rubber, measuring 6.5 inches in diameter (165mm) and weighing 15lbs (6.8kg) hitting the figure. Understandably, the *Spirit of Ecstasy* became one of the most desirable items for trophy hunters. Many mascots were lost which led to Rolls-Royce owners either having to alarm them or remove them altogether when parking the car.

In order to appear perfectly rectilinear each surface and component that made up the radiator shell had to have a slight curve; a practice appreciated by the Greeks who knew it as *entasis*. Look very carefully at the radiator and it will be seen that it's not entirely vertical but has the slightest forward inclination. Notice the vertical bars on the radiator - these should never be referred to as vanes but shutters. No longer do these open and close to control the flow of air; technology has come up with far more sophisticated ways to control cooling.

Care and fastidious attention to detail did not begin and end with the radiator. On every wheel fitted to the Silver Shadow there could be found a white spot which indicated that particular wheel had been carefully checked to ensure proper welding, finish and accuracy. Production checks as stringent as this were not good enough, however; one in every 1500 wheels was subjected to close examination and had a section cut out of it. The weld structure was inspected under a microscope and should the wheel have shown a fault the whole batch

was rejected.

Underneath a new Silver Shadow or T-Series Bentley it would have been possible to see 86 yellow-painted nuts (in fact, 91 but 5 were out of sight). Only when the torque of each nut had been checked and every component examined as far as was humanly possible was the dab of paint applied.

The same exacting tests were also applied to engine building. Cast iron cylinder liners were honed to an almost perfect roundness of within 0.0005 inches (0.0127mm). Crankshafts were machined to a tolerance of 0.0001 inches (0.00254mm) and connecting rods were weighed, matched and tested for any imperfection. Pistons, too, were selected in pairs and these were carefully checked for weight before being matched to make a complete set. All engines were run-in using gas on the test bench to the equivalent of 150 miles (240kms) and in addition one in every 100 was run for eight hours on petrol, after which it was dismantled and minutely examined. As if this were not enough, test cars were subjected to gruelling 50,000 mile (80,000kms) ordeals; the only time the cars were allowed to stop was for a change of driver and to refuel.

Low profile tubeless crossply tyres, 8.45 x15, on 6 inch (152mm) rim pressed steel disc wheels with five studs were initially fitted to the Silver Shadow and T-Series cars. Tyres by three manufacturers were specified: Avon, Firestone and Dunlop. Radial tyres were, of course, quite commonplace at the time, Michelin having introduced its 'X' tyre in the early fifties, but the

common feature of this type of tyre as opposed to crossplys was increased road noise. The first British car designed to use radial tyres, incidentally, was the Rover 2000, which made its debut in 1963. Although radial tyres were not adopted by Rolls-Royce until the early seventies, owners of earlier Silver Shadows and Bentleys have now fitted them quite happily and without detriment.

As can be expected, Silver Shadows and Bentleys were supplied with an impressive array of features such as two-speed self-parking windscreen wipers, screen washers, an alternator instead of a dynamo, laminated screen and reversing lamps, most of which would have either been optional extras or, more likely, quite unobtainable on some more humble cars. These features were in addition to the electrically-operated front seats, with their eight-way adjustment controlled by switches positioned centrally on the transmission tunnel, and, along with the electric windows, there was a facia panel rheostat, hazard warning lamps and a four-speaker radio (usually a Motorola). Even for a standard car as well-equipped as this there were optional extras such as seat belts. Safety harness anchorages were provided as standard but, at the time, the compulsory wearing of seat belts was a long way off.

The facia and instrument panel was a model of design ingenuity. Through the two-spoke steering wheel the driver had an excellent view of the main controls, traditional in appearance and neatly positioned. The instrument layout included speedometer, fuel gauge, ammeter and temperature gauge. Every eventuality had been thought of: the engine could not be started if either of the parking lamps were lit; warning indicators showed whether the hydraulic system was operating at the correct pressure and the driver could check the engine oil level without even having to leave his seat by pressing a button inside the car. In the event of a stop lamp failing a light on the facia indicated as much and for convenience the fuse box was positioned under the facia instead of being buried in the engine compartment. A later innovation, around the early to mid-1970s, was that the 'park' position was automatically selected if the ignition key was removed from its Yale-type lock without the parking brake first being applied.

One criticism did exist, though, which concerned the design and location of the electronic gear selector switch on the right-hand side of the steering column. As it was identical to the turn indicator, which was similarly positioned on the opposite side of the steering wheel, confusion between the two was inevitable with the possibility of disastrous results. The problem was quickly dealt with and the gear selector stalk replaced by a much larger lever with a positive upward-facing position.

Luxury and elegance

The interior appointment of the Silver Shadow was without compromise and the same careful attention to detail was applied to the Bentley T version.

Rolls-Royce went to great lengths to ensure hide and veneer quality. Each car was equipped with eight perfectly matched hides and each year a visit was made to Italy to buy only the best veneers. (Author's collection)

From the durable and washable headlining - Rolls-Royce publicists thought this important enough for customers to be made aware of it - to the Wilton carpet on the floor, the finish was the ultimate in good taste. It goes without saying that only specially selected top grade English hides were used for the upholstery and that each veneer for the facia and other areas where burr walnut was used, such as garnish rails, was the finest that could be found.

Front and rear seats were provided with centrally-positioned folding arm rests, the top of the rear rest lifting up to reveal a trinket box. For rear passengers there were wedge-shaped foot rests; cigar lighters were installed in each of the rear doors as well as on the facia and, built into the roof of the cabin, were four courtesy lights, the two at the rear incorporating reading lamps. Rear seat passengers were cossetted and for their convenience each had a map pocket and mirror. The luxurious interior would not have been complete without the usual folding picnic tables, finished, of course, in the same burr walnut veneer. Leather for the upholstery, padded capping rails and carpet bindings was supplied by Connolly Brothers who selected only the finest hides. In its raw state the material arrived at Connolly's premises in stiff bales, having been preserved in salt, but it was slender, finely grained and supple consignments of eight perfectly matched hides that eventually were supplied to Rolls-Royce. Available in eight colours - Beige, Tan, Grey, Blue, Red, Green, Scarlet and Black, each consignment was enough to upholster a complete car. In the trimming department skilled hands deftly cut and sewed the hides, forming them carefully over the seat frames to produce the ultimate armchair comfort. Good taste extended to the Wilton carpets, which matched the colour of the hide upholstery, and West of England cloth headlining which was optional to Ambla, a PVC material normally used; even lambswool rugs were laid in the rear passenger compartment.

The use of burr walnut in a Rolls-Royce and Bentley may be somewhat taken for granted but its selection and application was a highly skilled process, a work of art, in fact. Each year a visit was made to Milan in Italy to select the veneers and, for Micky Glynn, Rolls-Royce veneer expert, this was an annual pilgrimage. The amount of veneer required by Rolls-Royce each year was huge - enough to cover two football pitches - and it was not from a single supplier that the material was purchased. Visits were made to countless dank and dimly lit storerooms, often subterranean, where the veneers were closely examined; it was only those with the finest grain that were purchased. When the veneers arrived at Crewe the material was stored in an old air raid shelter, chosen because the underground conditions exactly matched those of the supplier. In selecting the veneers Micky Glynn was not looking for perfection; on the contrary, as the walnut tree grows so the grain of the wood around the foot of some of the trees is broken by a growth of warts and it is these imperfections, when the timber is sliced into wafer-thin sections, that produce the fine patterns.

The art in preparing the burr walnut is the careful cutting and matching of each sheet so that a veneer on one side of the car has a perfect mirror-finish on the opposite side. Up to eight veneers, all skilfully and invisibly joined, formed the facia panel alone and no two cars shared the same grain pattern. The glass-like finish on the veneers was the result of hours of painstakingly polishing and lacquering of the wood to bring out the full beauty of the grain. Surprisingly, approximately 60 per cent of each veneer was discarded during preparation and a further 10 per cent kept back for future use, should it be required, for example, to repair damage. The section of veneer that was retained was numbered and catalogued with the car's chassis details and stored carefully away in the dank conditions of the air raid shelter.

Performance - how it compared

The Silver Shadow - or any other Rolls-Royce or Bentley for that matter - wasn't bought for its fuel consumption. Initial tests showed that fast continental touring could consume petrol at the rate of 11 miles per gallon (25.7 litres / 100km) but a lighter touch on the accelerator could improve the figure to 15mpg (19lts / 100km). Overall, the Silver Shadow owner had to be content with something like 12mpg (23lts / 100km).

Most owners were happy to believe their cars without rival - but what of other large luxury cars that were available? Certainly large American machines such as the Buick Riviera and Cadillac Fleetwood were slightly less economical but who, at that time, worried about fuel consumption in the United States? The huge Mercedes 600 with its 6.3-litre V8 engine had tantalising features, which included pneumatic suspension and compressed air braking, which were almost as impressive as those of the Rolls-Royce. It equalled the Silver Shadow's fuel consumption but appalling handling let the car down badly. The clear winner in the fuel league was the Daimler Majestic Major, which was not as superior as a Rolls-Royce or Bentley, claiming an overall consumption of 15mpg (19lts/100km).

Of their four contemporaries the Rolls-Royce and Bentley were the slowest: as a maximum even the Daimler achieved a mean top speed of 120mph (192km/h) compared to 115mph (184km/h) for the Silver Shadow; the Mercedes outshone everything else with 125mph (200km/h) whilst the Buick and Cadillac settled for a more conservative 123mph (197km/h) and 118mph (189km/h) respectively. A closer rival, possibly, was the Jaguar Mark X which, by the mid-sixties, was powered by a 4.2-litre version of the trusty XK engine. This was a car much appreciated by many customers seeking the opulence of hand-crafted veneers and leather with refined engineering. At £2200, the big Jaguar offered good value for money with performance to match: maximum speed was 118mph (189km/h) but the trade-off was fuel consumption of 14 mpg (20lts/100km). The Daimler Sovereign, too, eventually became a contender in the rival stakes, especially in its Double-Six vogue. The new generation Daimler Limousine, which was based upon the floorpan of the Mark 10 Jaguar and powered initially by the 4.2-litre XK engine, might have been a more serious rival. In the event its role was more that of courtesy car, offering comfortable travel to and from airports and the like.

Through the gears and from standing starts the Bentley T and Silver Shadow were never clear winners, but after all who bought the cars for sheer performance? As far as devotees were concerned, the vehicles offered superlative comfort with more than enough speed and prices were academic. The Daimler cost a mere £2749 but failed to achieve the same degree of respect; the price tag of the Mercedes was a massive £9000 less a few pennies, and the American offerings could be bought for between £4400 and £5200. At around £6500 the Silver Shadow and Bentley T oozed quality and refinement, dignity and sophistication. These cars were not merely assembled but created, products of dedicated engineering and skilful hands. Perhaps more importantly than anything else, they epitomised a deep-seated loyalty to the company and its traditions.

Not just new models from the most respected car manufacturer in the world, the Silver Shadow and Bentley T were the beginning of a new era. The cars that followed included a second series Silver Shadow, the Shadow II and, alongside it, the Bentley T2; coachbuilt versions were offered and these included the now rare James Young-bodied two-door saloon and the delightful Corniche from Mulliner, Park Ward. The Corniche, incidentally, remained in production long after the Silver Shadow had been superseded by the Silver Spirit, and it was almost thirty years after announcement of the Silver Shadow that the last car was built. Long wheelbase models, which eventually evolved as the Silver Wraith II, and the controversial Pininfarina Camargue, all made their debut within the lifetime of the Silver Shadow and are examined later in this book.

III

WHISPERING POWER

The mid-sixties was a time of both reward and disappointment at Rolls-Royce: reward because of the achievements which culminated in production of a car which, for Rolls-Royce, was quite new in its engineering concept, and disappointment at the level of losses sustained in the motor car division. It has to be said, however, that losses during this period had been fully anticipated because of the enormous costs involved in development of the Silver Shadow, and that the board of directors, who had been kept fully informed at all times, had given its approval in this respect.

As the Silver Shadow and its Bentley equivalent went into production, the cars' future was overseen by the company's general manager, Ray Dorey OBE (*Dor*). Dorey had accepted the position of GM at Crewe in the early fifties, rather than head the company's interests in Canada, a position he had also been offered. His career with Rolls-Royce began in the late twenties after he graduated from university. Dorey was invited to join the experimental division at Derby, working on aero engines, and immediately made a name for himself testing engines on the run-up to the 1929 Schneider Trophy Race. Cars and motoring were amongst his keener interests and during those early

Many hours were spent perfecting the frontal appearance of the Silver Shadow and its Bentley equivalent. John Blatchley's styling ensured careful attention to detail, such as discreet positioning of the Rolls-Royce emblem. T-Series cars carried the Bentley emblem, of course. (Courtesy Rolls-Royce Enthusiasts' Club)

This fleet of Silver Shadows makes an interesting spectacle, but for Rolls-Royce the change in production methods necessary to build the car meant a complete revision of the company's time-honoured traditions. This was the largest single order from a customer, the Peninsular Hotel, Hong Kong. (Author's collection)

years at Derby he owned a Bugatti (as a point of interest it was Ettore Bugatti who had referred to W.O. Bentley's cars as the fastest lorries on earth!)

During his term as general manager Roy Dorey witnessed a frustrating period in Rolls-Royce's finances. Just as the motor industry recovered from the effects of shortages caused by the aftermath of war and was able to look forward to a new decade - the sixties - with confidence, severe restrictions were placed upon the industry by the then chancellor of the exchequer, Selwyn Lloyd. In his budget, the chancellor created taxes that would deter potential customers from buying expensive motor cars. Dorey had no recourse other than to cancel orders with Rolls-Royce suppliers in order to reduce the effect of a sudden downturn in sales. The motor industry suffered as a whole, of course; Rolls-Royce was not the only company manufacturing highly-priced cars. Public perception of large-engined cars, such as the Silver Cloud and Bentley S-Series, was badly affected by the chancellor's tax burden and it was upon a campaign of damage-limitation that the general manager embarked. It was to this end that two interesting managerial decisions were taken at Rolls-Royce: firstly, the ultimately abortive discussions with BMC took place and, secondly, a radically different Silver Cloud and Bentley were unveiled - the Series III and S3 models.

Although potentially a successful venture, the collaboration with BMC turned out to be a dismal failure. There were few customers for the Vanden

Plas Princess R and a Rolls-Royce-engined Austin Healey never materialised. Rolls-Royce was left with two engines which Jack Phillips, who had been responsible for development of the V8, had purposely designed: the 3.9-litre F60, which produced 178hp at 4850rpm and was fitted to the Princess 4-litre R, and the G60, which was of the same capacity but produced 268hp at 6000rpm. It was the latter engine which was proposed for the Big Healey. With sales of the Princess R at a disappointingly low level, the F60's production line at Crewe remained idle for much of the time before being dismantled. Luckily, both engines have been saved for posterity at the Rolls-Royce Enthusiasts' Club headquarters at Paulerspury.

Losses in the motor car division escalated rapidly: a deficit of £300,000 was recorded in 1963 and within two years this amount had trebled. A year later, 1966, the loss increased further to over £1 million and it was not until two years later that a profit - albeit a small one - was made. Ray Dorey had foreseen, in the mid-fifties, that developing a car as complex and innovative as the Silver Shadow would present a difficult period for Rolls-Royce and, as a precaution, had strengthened the design department with some very able stylists and engineers. His philosophy

had worked and, as has been revealed in earlier chapters, the department produced some outstanding results. Having successfully overseen the launch of the Silver Shadow and Bentley T-Series, Dorey decided to retire in 1968. As he had predicted, the gestation period of the models had proved very challenging and, what was worse, he had found himself caught in the crossfire between some very strong personalities. Harry Grylls, who had virtually single-handedly engineered the Silver Shadow concept, appreciated the importance of such an innovative design, while Geoffrey Fawn (*GF*), who had joined the aero engine division of Rolls-Royce at Derby from the nuclear experimental establishment at Harwell, was critical of a car that incorporated so many new ideas at one time. The friction between Dorey and Fawn was often very evident and the final two years of Dorey's career were somewhat turbulent.

Geoffrey Fawn succeeded Ray Dorey and, at about the same time, Harry Grylls also took retirement. The position of chief engineer was taken by John Hollings who remained in that post until 1981 when he was appointed technical director. Hollings was no stranger to Rolls-Royce, having joined the company in 1948 as an aero engine designer. In 1957 he was appointed

chief designer on nuclear reactors for submarines and, in 1962, moved to Dounreary in Scotland as manager and chief engineer at Admiralty Reactor Test Establishment. A further move, in 1965, to take up the post of chief quality engineer at the car division of Rolls-Royce took John Hollings to Crewe.

Taking over from Harry Grylls would have been a difficult task for any successor to the post of chief engineer and Hollings, having acknowledged his predecessor's enormous contribution to both the auto industry and the motor division of Rolls-Royce, set about building upon the legacy that had been left. When talking to John Hollings about his early days as chief engineer, he recalls he knew less about motor cars than the engineers under his control. Valuing them greatly, he therefore concentrated upon managing his team effectively, leading them in the right direction. John Hollings accepted that his task was one of continuation and improvement of the Silver Shadow throughout the remainder of the car's production period, which he understood needed to be somewhat longer than previous models due to the car's extraordinarily high development costs. There nonetheless had to exist a plan for the car's eventual replacement which, as history has shown, evolved as the Silver Spirit, and initial work on the car, designated SZ, began in 1969.

The period throughout the late sixties and the seventies, however, was particularly testing for the new chief engineer who had to oversee a number of important events. Firstly, the American motor industry was in a state of upheaval as a result of US safety and emission control regulations. As a company that relied heavily on selling cars into America, Rolls-Royce was forced to comply with the legislative changes or risk losing this lucrative market. Secondly, there was the fuel crisis which erupted in the early seventies, the shock waves of which were felt worldwide.

While this was happening continual progression and development of the Silver Shadow and T-Series Bentley led towards the introduction of the second series models in 1977. A third development was the introduction of a special-bodied derivative of the Silver Shadow, the Camargue coupé. This was intended as the marque's flagship and an alternative to the already prestigious two-door saloon which was renamed the Corniche (the car proved somewhat controversial, but more of this later). The most devastating event of the period was undoubtedly the financial collapse of Rolls-Royce on February 4th 1971. On what became known within the company as Black Thursday, Rolls-Royce - national institution and possibly the most marketable name in the world - teetered on the edge of disaster.

The seventies: a decade of engineering changes

One of the first engineering policy decisions made under John Hollings' control was that concerning the coachbuilt cars produced by Mulliner, Park Ward. Although the coachbuilt models based upon the Silver Shadow and T-Series Bentley are described in greater detail in a separate chapter, it is necessary to emphasise it was the intention that these cars should pioneer any engineering modifications before they were applied to the usual production, or 'Crewe', cars. In addition to the very elite two-door model - later known as the Corniche - it was decided to produce a wholly new coupé which supported styling from outside the company, rather than from within Crewe's own styling department. Italy had a reputation for producing some of the world's most beautiful cars and it was the stylist Pininfarina whom Rolls-Royce approached.

Pininfarina had been responsible for a number of designs, including a fastback version of the Silver Dawn which had been exhibited at the 1951 Turin Motor Show. The company had also produced five cars based upon the Bentley Mark VI chassis as well as a single example of the R-Type Continental. Pininfarina also produced for Lord Hanson a single coupé based upon the T-Series Bentley and it is generally considered that it was this car which led to the Italian styling house being commissioned to design the Camargue. Nearly thirty years later enthusiasts may wonder why Rolls-Royce decided not to entrust the coupé to its own stylists, especially when

Rarely have so many Silver Shadows been seen together! (Author's collection)

considering that the department had produced such exciting machines as the Bentley R-Type and S-Type Continentals under the direction of Ivan Evernden and John Blatchley. It has to be said that within Rolls-Royce circles the decision to place the Camargue, as the coupé eventually became known, outside the company was met with some misgivings! The idea of a car such as the Camargue had originally been proposed by Geoffrey Fawn who had the notion that Rolls-Royce needed an exclusive model with an exotic sounding name.

In accordance with the policy of modifying the Rolls-Royce 'Crewe' cars, changes were therefore specified for the Camargue a year before being featured on the Corniche. A year after that, revisions were finally applied to the production models, the Silver Shadow and T-Series saloons. The reason for this tiered system was that new features should initially appear on the most expensive cars to satisfy first the company's most wealthy customers.

Of strong influence during the

seventies was the need to meet the stringent US regulations aimed at improving safety and control of exhaust emissions. Amongst the first tasks to which the development department had to attend was that of head impact, which meant conducting tests using a 6.5 inch (165mm) diameter sphere to simulate a driver's or passenger's head striking the facia. Similarly, tests were carried out imitating a rear passenger's head hitting the top of the rear section of the front seats. Some of the modifications which resulted affected the Silver Shadow facia, which was redesigned to prevent a human head from striking any control knobs or switches on the dashboard.

In some cases, where the roof had to provide enough strength to absorb sudden impact, the design of the Silver Shadow exceeded all requirements and no modifications were needed. This ably demonstrated the safety margins built into the car, especially as the roof was required to withstand 5000lbs (2273kg) without deflecting more than 5 inches (127mm).

British or European car manufacturers like Rolls-Royce, who relied upon American sales, could not afford any complacency when complying with Federal Motor Vehicle Safety Standards. While having to conform to regulations was obviously very expensive and time-consuming, there were, however, important benefits and it was these that helped place Rolls-Royce safety specifications far ahead of most other car makers. In general terms, US safety legislation was ten years in front of Britain's and, allowing for some flexibility on America's part, it enabled Rolls-Royce to keep something like seven years ahead of other British car manufacturers. A modern example is airbags became available on most British and European cars in the early nineties. Rolls-Royce, however, had this technology by 1984.

Side impact protection was an important issue for Rolls-Royce in the seventies when American standards required an intrusion test on each door. The aim, of course, was to deflect two cars meeting at an angle and prevent them interlocking and spinning round, thereby reducing the risk of injury. The tests which the development department had to instigate were, by necessity, complex, but it did mean that special bars were fitted to the insides of the doors, forming a barrier between the locks and hinges. An outcome of the tests was a modification which actually weakened the wooden waist-rail finishers in each car so in the event of an accident they would not splinter and spear the occupants.

Front and rear impact tests - as

66

previously described - were a necessity and these, generally, were carried out under the supervision of Jock Knight who regularly managed to 'crash' each test car at least four times before it had to be scrapped. An important modification was a new type of bumper required by US regulations which came into force in September 1972. The bumper, *in situ* on the car, had to withstand two tests, the first of which involved the car being crashed into a solid wall at 5mph (8km/h). To meet this requirement the bumpers were mounted upon specially designed shock absorbers filled with silicone rubber, which gave a better performance than an oil-filled damper. After impact, the energy absorbing bumpers had to return slowly to their original positions without any damage or deformation. The second test involved the bumper being hit at an angle at 3mph (5km/h) by a weight equivalent to that of the car to which it was fitted. This ensured the bumper beam had adequate built-in strength to withstand minor impacts.

Along with meeting the stream of safety demands which emanated from America, Rolls-Royce had to conform to exhaust emission controls which were being instigated by a number of countries in addition to the state of California in the US, which was the most stringent. To satisfy the regulations, progressive changes to the Silver Shadow were necessary at the rate of two year intervals. Not only did Rolls-Royce cars suffer from reduced fuel economy as a result of these regulations, but performance was also affected. The need to run engines on leaner fuel mixtures resulted in an overall loss of power and this was a major contributing factor in the development of a new V8 engine in 1970.

By lengthening the stroke and increasing capacity of the existing V8 engine from 6230cc to 6750cc, Rolls-Royce effectively overcame the power loss problem. Needless to say, the company declined, as usual, to announce its bhp rating although estimates calculated it at approximately 220bhp, a possible 20bhp increase over that of the 6.23-litre engine. Further controls followed which curbed the use of 5-star, 99 octane fuel in favour of 4-star, 97 octane fuel, and thereafter work was carried out in preparation of 91 octane lead-free fuel. In America, serious efforts to reduce pollution levels were being made during the mid-seventies, which resulted in the introduction of catalytic converters.

Following the fuel crisis of the early seventies, when petrol rationing in the UK was all but a reality, Rolls-Royce, together with the motor industry as a whole, regarded the question of fuel economy with some priority. John Hollings acknowledged the importance of the issue when he presented the autumn lecture on the engineering of Rolls-Royce cars in 1982. Hollings told his audience: *"The two most important factors are vehicle weight and engine efficiency. We are working on the reduction of vehicle weight by painstaking attention to detail and we are working on improvements to the engine which will increase efficiency by allowing higher compression ratios without detonation. We shall also reduce the engine capacity and improve breathing to restore power. In this way we expect to provide a significant improvement in fuel economy in a few years' time using our present V8 engine with some significant changes."*

American car manufacturers were given very precise time limits in which to comply to the US Federal safety and emission control regulations. Luckily, Rolls-Royce, as a foreign producer, was allowed some leeway by the American government in implementing the requirements; had this not been the case Rolls-Royce would have lost important export orders, resulting in almost certain disaster.

Throughout the Silver Shadow's era it is now evident that the Bentley marque was allowed to decline to such an extent that it almost disappeared entirely. Happily, this did not happen and its importance was acknowledged with new generation models such as the Bentley Eight, Mulsanne, Brooklands and Continental models. Today, the future for Bentley-badged cars looks far more secure, a situation acknowledged by the unveiling of the Java Concept car in the mid-nineties. In what was recognised as the ultimate in badge-engineering, the T-Series Bentley was completely overshadowed by its Rolls-Royce counterpart and in some of the company's publicity brochures received hardly any mention.

The Bentley's role at the time was considered by Rolls-Royce management to be subordinate; although the T-Series cars were Silver Shadows except in name, badge and radiator shell,

Bentley-badged cars were built in remarkably low numbers in comparison to the Silver Shadow. The difference in cost between the two models was so little that most customers opted for the prestige of a Rolls-Royce radiator. There was always some demand for the Bentley T-Series cars, however, and the two seen here make an elegant pair.
(Courtesy Rolls-Royce Motor Cars Ltd.)

they were nonetheless denied that certain reverence. In certain companies a Rolls-Royce might be awarded to the chairman, while the deputy chairman would have the Bentley, seen as not having quite the same status. In one instance of price structuring, the Bentley version was forgotten altogether, hence the nominal difference of just a few pounds when the error was realised! The fact that the Rolls-Royce-badged car was just £64.00 more than the T-Series Bentley obviously decided which would be the more popular buy.

Describing his involvement with the Silver Shadow, John Hollings re-

called some of the early problems experienced with the car. The ride and handling did have some shortcomings which, to a great extent, were rectified when the self-levelling suspension was removed from the front axle; the fitting of compliant suspension to the front wheels, which enabled the use of radial tyres, helped even further. The car's electrical system was often a cause for concern and, more often than not, troubles were traced to loose connections. The wiring loom which - on a car of such complexity - was by essence a complicated affair; behind the facia alone there were something like 300

cables. Close collaboration was essential with manufacturers such as Lucas, but this was not always entirely satisfactory. Noise from the differential was often a matter of complaint and hundreds of hours were spent trying to cure the problem. Ultimately a dampening device was installed between the propeller shaft and flange on the differential casing.

There were fewer problems with the saloons once the Camargue and Corniche variants were introduced, for the simple reason that any irregularities had mostly been eliminated on the coachbuilt cars within the two years it

had taken to apply the same modifications to production models. It was for this reason that John Hollings regarded the Series II cars, which were introduced for 1977, as being very satisfactory and, from his point of view, the late seventies and early eighties were a period of fulfilment and reward.

Shadow chronology

Before reviewing the changes that were made to the Bentley T and Silver Shadow during production, it is important to explain the notation system used to denote chassis identification.

A typical chassis number would consist of three prefix letters, followed by up to five digits, such as SRX 12345 or LBH 23456. The first letter establishes the exact body style: S - standard saloon; L - long wheelbase saloon; C - two-door saloon and convertible up to chassis number 6632; D - convertible from chassis 6646; J - Camargue. The second letter confirms whether the car is a Bentley (B) or Rolls-Royce (R). Thirdly, the prefix determines right-hand drive (H) or left-hand drive (X) and, in the case of the latter, should the car be destined for North America, the X will be replaced by a letter: A-G or J-L dependent upon model year as A=1972, B=1973, C=1974, D=1975, E=1976, F=1977, G=1978, J=1979, K=1980 and L=1981. The digits that followed the letters simply verified the chassis numbers which were in batches: 1001-4548; 5001-5603; 6001-8861; 9001-26708 and 30001-41648 (Series II models).

Rolls-Royce adopted a policy of

continual improvement throughout the entire lifespan of the Silver Shadow and Bentley T-Series cars and, by studying the list of specification changes that were applied to these models, it can be appreciated just how much development work was carried out. Many of the changes, however, while important in maintaining the company's upgrading policy, were, to a major extent, quite insignificant to the owner. Within a few months of its launch, the first derivatives of the Silver Shadow appeared: the two-door saloon by James Young in January 1966 and, two months later, the H.J. Mulliner, Park Ward two-door saloon. These and other variants are described in-depth in the next chapter.

Amongst the first changes to the saloon cars, from chassis number 1467, was the provision of a lighter brake pedal movement at the end of 1965. From October 1967, and chassis number 3000, a Saginaw power steering pump replaced the Hobourn Eaton type, therefore complementing Saginaw's own recirculating ball steering system. A new type of boot lid seal was fitted from chassis number 3367 and the opening quarter light fitted to the front doors was replaced by a fixed type. During 1968 there were a number of important changes: a revised handbrake was fitted from chassis 4469 and from chassis 6300 new windscreen

washers were specified. A higher steering ratio applied from chassis 6429 and road wheels were provided with a flat ledge rim from 6771. A further change to the steering was made from the end of 1969 when a smaller steering wheel, 16 inch (410mm), was fitted from chassis 8222.

By far the most important specification changes during 1968 were those concerning the automatic gearbox and modification of the front anti-roll bar as well as fitment of an anti-roll bar at the rear.

All cars, left- and right-hand drive, were fitted with General Motors GM400 transmission from chassis 4483. Although the GM400 gearbox was bought-in direct from America, the electric selector actuation was not part of its specification and this, therefore, was added at Crewe. Rolls-Royce was the only manufacturer to fit electric actuation to this type of gearbox - which allowed the lightest finger-tip control - and whilst such cars as the Cadillac were fitted with the same unit, ratio selection was operated manually. Originally fitted to left-hand drive cars only, the GM400 gearbox, when fitted to home market cars, had the following ratios: 1:1 (top), 1.5:1 (intermediate), 2.5:1 (first) and 2:1 (reverse).

Changes to the suspension, which include a strengthened front anti-roll bar and the addition of an anti-roll

After 1970 all cars were fitted with the 6.75-litre engine. The car pictured dates from 1972 and is built on chassis SRH 13474. Note the ventilated wheel discs. The location, by the way, is Castle Ashby, Northamptonshire. (Courtesy Rolls-Royce Enthusiasts' Club)

device at the rear, apply from chassis 4528. Importantly, this modification applied to British and European market models only which demanded a firmer ride. For the American market, which preferred a softer, more wallowing ride, these changes were not implemented.

Between August and December 1969 several specification changes were announced: longer bonnet locating dowels were fitted from 7209; alternators (C.A.V. type) were made standard from chassis 7500 and from 7620 stainless steel exhausts were fitted. It was found necessary to strengthen the handbrake calipers and modify the design of the hazard warning switch from chassis numbers 7650 and 7904 respectively. A major revision to the suspension system, at chassis 7404, was implemented when the forward height control was deleted. The reason for this was twofold: firstly, the front self-levelling device had little or no work to do and, secondly, its removal improved the handling and reduced some of the vagueness in the steering which had become a matter of criticism.

As well as alternators being standardised throughout all model production, chassis number 7500 marks the point where air-conditioning was specified for all cars. Before this, air-conditioning had been an optional extra and only about half the cars sold had it fitted.

A combination of continual specification improvements and tight controls from the USA concerning exhaust emissions were having a serious drain-

ing affect upon engine power. This situation, however, helped influence one of the most important engineering decisions in the history of the Silver Shadow and its derivatives. In order to correct the situation with particular emphasis on improving torque, introduction of the 6.75-litre engine occurred at the end of 1969 in readiness for the 1970 model year. America, in fact, was not the only country concerned about pollution issues as Ja-

pan and, perhaps surprisingly, Australia were also legislating to fight the problem.

John Bolster, testing the 6.75-litre-engined Silver Shadow for *Autosport* in December 1970, was impressed at how much low-speed torque had been improved. Overall speed also increased and he found the car easily achieved 118mph (189km/h). Taking advantage of a long straight stretch of road he was able to better that by at

The facia depicted here was designed to comply with American safety legislation. Note the thickly padded edges which were intended to prevent injury in the event of an accident. Stylist Martin Bourne remembers using a wooden ball, the size of a baby's head, to ensure that switches and instruments did not protrude too far. (Author's collection)

1   Hand brake / stop lamp bulb failure warning lamp
2   Gearchange selector lever
3   Fuel / oil level indicator and warning lamps test switch
4   Adjustable outlet and control for fresh or cold air
5   Windscreen wiper / washer switch
6   Accumulator warning lamps
7   Loudspeaker balance control
8   Adjustment knob for hands
9   Aerial switch
10 Fuel level warning lamp
11 Instrument lamps switch
12 Coolant level warning lamp
13 Adjustable outlet and control for fresh or cold air
14 Cubby box lock
15 Map lamp switch
16 Air-conditioning outlet
17 Lower heating and ventilation switch
18 Rear window demister switch
19 Blower motors switch
20 Ash tray
21 Front seat switches
22 Cigar lighter
23 Hazard warning switch
24 Upper heating, ventilation and refrigeration switch
25 Control for side scuttle wall outlet

least another couple of miles per hour. As has already been discussed, the extra cubic centimetres were derived from a redesigned crankshaft which produced a lengthened stroke. 6230cc was stretched to 6750cc and the power output, not disclosed, of course, increased to somewhere around 220bhp. Rolls-Royce, incidentally, was carrying out a lot of experimental work on engines at this time and was even considering a power unit of over 7 litres. By increasing the stroke to 4.2 inches (103mm) it had been possible to enlarge the existing V8 to 7269cc and a prototype engine was fitted to the last purpose-built experimental car. Registered XMA 66M, the car was extensively tested in Europe and, on one occasion, with John Gaskell at the wheel, it achieved 127mph (203km/h). Although performance would have been greatly improved, fuel consumption would not necessarily have suffered as a result, although this fact was not appreciated at the time by a motor industry preoccupied with the world fuel crisis. Ultimately, it was found that a similar performance could be achieved by restricting the engine size to 6.75 litres.

During 1970 the engine underwent further modification from chassis 9900, when a redesigned camshaft was fitted. This was introduced to amend the valve timing on the Corniche to give increased power output. The modification was not applied to the Silver Shadow saloon, however, until a couple of years later. There were a number of other modifications during the year, too: new brake calipers were specified at chassis 9380 and the suspension height control sensitivity mechanism was adjusted from 9393. Centralised door locking became a feature of the cars and was fitted from chassis 9658. The front seat mechanism received attention for the 1971 model year (at chassis 9630) and a Trico windscreen wash system was specified at 9898.

American safety legislation was responsible for a new-look facia which was also redesigned to incorporate a central console. All dashboard edges were softened by the application of thick padding, which gave the facia a narrower appearance. To comply with US regulations it had been necessary to restyle the dashboard and this was achieved in a most bizarre way. Martin Bourne recalls having to check the distances between the upper and lower padded surfaces with a wooden ball - intended to represent a baby's head - to ensure protection from protruding switches.

The earlier facia, affectionately known by company employees as the Chippendale, gave way to a more aesthetically pleasing design, although enthusiasts of pre-1970 cars may disagree ... Instrument layout was improved and installation of the clock in the centre of the facia, instead of left of centre on right-hand drive cars, gave a more balanced look. Switches and controls were grouped together in a neater style than previously and warning lamps installed in separate panels. The central console, as well as housing the air-conditioning outlet and heating and ventilation controls, also accommodated the radio, seat adjustment switches, hazard warning switch, cigar lighter and ashtray.

Even as late as 1970, Rolls-Royce was still experimenting with different types of self-adjusting suspension. Tests were carried out using both

A proud owner (whose identity, unfortunately, is unknown) pictured with his Rolls-Royce, the 10,000th Silver Shadow built. (Courtesy Rolls-Royce Motor Cars Ltd.)

Citroën and Rolls-Royce accumulators and, talking with John Hollings, it appears the French product was substantially more reliable. Use was also made of a Telegas self-levelling system devised by Moulton, inventor of the bicycle of the same name, but this proved unsuccessful. The design was something of a hybrid and employed a hydrolastic layout similar to that devised for the British Motor Corporation but including Citroën type spheres. Although experiments lasted for some months, it was eventually discarded and the car to which it was fitted was converted back to the usual suspension.

From chassis number 10,000, the approximate delivery date being January 1971, speed control became an optional feature - at chassis 10322 - and a windscreen wash-wipe system was introduced from 10400; ventilated front wheel discs, to improve overall braking performance and avoid fading, were fitted from chassis 10500 and by the middle of the year there was improved radio suppression and modification to the ignition system.

The speed control, otherwise known as cruise control, enabled the driver to keep the car at a predetermined speed, whatever the road incline. On experimental and early production cars, the control device was operated from the facia but some years later was moved to the end of the gear selector lever. A safety device was built into the control which made it non-operational at speeds under 30mph (48km/h) or over 80mph (128km/h). In addition, it would instantly disengage as soon as the brake pedal was depressed.

After June 1971, Rolls-Royce gave further attention to the Silver Shadow's suspension system from chassis 11130 and changed the type of hydraulic fluid used. All Silver Shadows use a synthetic brake fluid of various specification, culminating in RR363, which attacks paintwork if allowed to come into contact with it. At about the same time, from chassis 11155, cars destined for the USA and Canada were fitted with modified brakes, which resulted in a slightly lighter pedal action, and, applying to all cars, the steering ratio was altered to 17.5:1 at chassis number 11501.

Another revision to the self-levelling suspension was made at chassis number 11970, this being the installation of a modified height control valve from approximately the end of the year. Further improvements followed in quick succession: a slightly redesigned rear seat was specified and Kangol seat belts were fitted as standard to all cars being exported to the USA; modified striker plates were fitted to the doors, and a new type of expansion tank was fitted to the car's cooling system. A different type of piston ring was fitted to the engine, from chassis 12657, and minor alterations - such as modification to the windscreen wiper and facia panel light switches - were also introduced.

Of significance were the modifications made to the chassis dimensions introduced from 1971. Although the changes were minimal they were nonetheless of extreme importance, being necessary for improved road holding and handling. An increase in the length of the wheelbase, from 119.5 inches (3035.3mm) to 120.062 inches (3049.6mm), and the track - from 57.5 inches (1460.5mm) to 59.5 inches (1511.3mm) at the front and 57.75 inches (1466.9mm) at the rear - was made in order that compliant suspension and, later, during 1972, radial ply tyres of 205VR x 15, could be accommodated.

Compliant suspension, the term used by Rolls-Royce for the revised suspension layout, involved the design of a new system of subframe mounting. The stainless-steel, wire-mesh type of dampers - more commonly referred to as pot scrapers or 'pan scrubbers' - were eliminated and in their place were fitted specially designed rubber bushes. A cranked-arm top link replaced the upper triangle levers fitted originally and a measure of fore and aft movement (graphically described as 'rock and roll') was obtained from a compression strut linked between the top link and the subframe. It has to be added that compliant suspension - fitted to all cars from chassis number 13485 - was first introduced on the Corniche models from chassis number 12734 with delivery of these cars being effected from the end of 1971 and beginning of 1972.

Black Thursday
Without any doubt whatsoever, 1971 was a devastating year, not only for the

Fritz Feller took over from John Blatchley as chief stylist in 1969. Like everybody at Rolls-Royce, he was stunned when the company went into receivership. (Courtesy Martin Bourne)

car division of Rolls-Royce, but the company as a whole. The aero division of Rolls-Royce, as well as the motor car division, had established itself as a national institution and the double 'R' symbol on the casing of the world's most respected engines instilled unparalleled pride.

Alas, the company had become embroiled in a costly issue over its new aircraft engine, the RB211, and for 1970 recorded a deficit of £3 million. The engine was being developed for the new generation of wide-bodied jets; in particular the Lockheed Tristar. Difficulty had been experienced with construction of the fan blades which, during testing, were found to be susceptible to bird strikes and it was necessary to form the 25 huge blades from titanium instead of the carbon fibre material originally specified. Not only was titanium very much more expensive, but development costs soared out of control as the engines failed to perform as well as expected, due to the increased weight of the blade material. (Titanium, incidentally, was also used to manufacture the turbine blades of Rover's gas turbine car JET 1, a project in which Rolls-Royce had once shared an interest).

Acknowledging the difficulties which faced the company, Rolls-Royce management took two courses of action. Firstly, Sir Stanley Hooker, the former chief engineer of the company's jet engine division, was consulted and agreed to return to Rolls-Royce from retirement to undertake the redesigning of the RB211, which he successfully accomplished. Secondly, the Brit-

ish government was approached for help; assistance was not forthcoming which left Rolls-Royce no alternative but to appoint a Receiver to take over the company's affairs.

John Hollings remembers clearly the 4th February 1971, which became known at Rolls-Royce as Black Thursday. It was appreciated by the car division, which was operating at a profit, that the aero division was in trouble but it was not known to what extent. At 9am John Hollings was called to the boardroom at Pym's Lane and given the news that E. Rupert Nicholson had been appointed Receiver. Hollings was devastated. (An interesting point is that Rolls-Royce was never declared 'bankrupt' as assets far exceeded debts or potential debts). Martin Bourne recalls the air of disbelief in the styling department when, shortly after 9am, Fritz Feller - who had taken over from John Blatchley as chief stylist - announced the news. Everybody was stunned: throughout Rolls-Royce there had always existed a unique sense of loyalty - that feeling is evident to this day - and it was felt that a tragedy had befallen the whole family that was Crewe.

Trading ceased at that moment; no deliveries could be made and supplier's lorries were turned away at the gate - including the brewer's dray with

the beer for Saturday night's dance in the ballroom. That really brought home the gravity of the situation! Leaving the boardroom a few minutes after being told of the bankruptcy, John Hollings' task was one of reassurance; he realised it was essential not to allow staff morale to suffer any more than it had already and therefore, as chief engineer, he toured all departments in order to try and defuse the situation. After that, as far as he was concerned, it was a matter of 'business as usual'.

And business as usual it was. The Receiver, fortunately for the car divi-

David Plastow, marketing director since 1967, was promoted to the position of managing director in 1971. (Courtesy Martin Bourne)

sion, was sympathetic and gave specific instructions that car output was not to be affected. In addition, accepting that the company was within five weeks of launching a new model, the Corniche - a modified version of the existing two-door saloons and convertibles, he agreed this should go ahead as planned. Geoffrey Fawn had already been recalled to Derby in January and David Plastow (*DP*) was appointed managing director, a promotion from the post of marketing director which he had held since 1967. These were challenging times for both Rolls-Royce and the new MD, and David Plastow worked in close association with Rupert Nicholson.

The collapse of the company also affected Rolls-Royce in America. Rolls-Royce Inc. became Rolls-Royce Motors Inc. and, because the new company was interested in motor vehicles only, instead of aero engines primarily and cars as an off-shoot, sales of vehicles increased dramatically. From 700 cars a year, sales increased to over 1000 cars per annum. As for the aero division of Rolls-Royce, this was nationalised and effectively separated from the car making business. Rolls-Royce Motors Ltd. was formed and, eventually, floated on the British stock market as a public company; David Plastow retained his position as managing director of the newly created concern and Ian Fraser was nominated chairman.

The launch of the Corniche went ahead as scheduled and, as it happened, the new car acted as something of a fillip to company morale. Unveiled at a special celebration in Nice on the

French Riviera, the party of motoring journalists invited to the ceremony might have wondered at the seemingly incongruous timing for such a spectacular occasion. Any such thoughts soon evaporated, however, when David Plastow explained that the entire event had been staged for less than half the cost of a Corniche. Had the model not been launched with such high profile there could have been at risk not only the status of Rolls-Royce Motors, but also the confidence of the motor industry in general - especially that of Rolls-Royce's component manufacturers and suppliers.

Car production throughout this turbulent period in Rolls-Royce history actually increased: in 1970 2009 cars of all types were built and by 1971 the figure had risen to 2280; 2470 cars sold were recorded for 1972 and, for 1973, 2760 orders had been fulfilled. Rolls-Royce order books were full and the future looked promising.

With Rolls-Royce cars re-established, work on progressive development continued: from chassis 13051 modifications to the distributor advance were made in respect of those engines with emission controls and the sound system was improved with Philips double-cone speakers from chassis 13178. The American emission control requirements were becoming very effective and to comply with the 1973 Detox regulations, the cars specified for the USA received specially modified engines from chassis number 14954. All cars received a modified hydraulic fluid reservoir at chassis 14980 and for the 1973 model

year cars exported to the USA had to be fitted with larger 'brakes on' stop lights. Another modification to the hydraulic system was the specification of Castrol fluid, RR363, effective from chassis 15638. The brakes received further attention later in 1973 when the pedal was given an even lighter feel (from chassis 15854) and all cars were fitted with a new type of disc brake pad, referred to as type M170, at chassis 15950.

For the Swedish market, cars from 1973 - at chassis 15855 - were fitted with a headlamp wash and wipe system. During the same year, USA and Canada-destined cars were supplied with energy absorbing bumpers at the front and rear; additionally, head restraints were incorporated into the rear seats and a remote control device fitted to enable the external mirrors on these cars to be adjusted from inside the car. As has already been described, the bumpers had specially designed shock absorbers built behind the stainless steel frames and were designed to return to their original shape after minor impact. The styling department at first was not impressed at the appearance of the US regulation bumper but, in time, not only came to appreciate it but actually advocated it for the Second Series cars. Marque enthusiasts had not been happy about them either, many complaining that the car's appearance was adversely affected. As it happened, of course, similar styles of bumper were subsequently fitted to the majority of cars and the issue was quickly forgotten. Look closely, however, at two cars, one with traditional

Cars specified for the American market were equipped with special impact absorbing bumpers as shown here. The car in the photograph is actually a Silver Shadow II. (Courtesy Rolls-Royce Motor Cars Ltd.)

bumpers, the other with energy-absorbent type, and it will be seen that the car with safety bumpers has a slightly restyled radiator shell, the bottom of which is shortened to accommodate the 2.5 inch (57mm) fore and aft movement of the bumper insert. Further observation will reveal that a shroud exists around the base of the bumper assembly extending upwards which led to the rectangular air intakes being deleted.

Along with the provision of radial tyres in 1972, some attention was given to facia design. Where there had been separate gauges for oil pressure and water temperature, warning lamps and buzzers were fitted instead. The switch box was moved from the middle of the dashboard to either the left-hand or right-hand side, depending on whether the car had left- or right-hand steering. The windscreen wash-wipe switch which, by this stage, had an intermittent wipe setting, was also relocated to the direction indicator stalk on the steering column and, in its place in the centre of the facia, were a cluster of warning lamps for low fuel, brake pressure and partial failure and hydraulic fluid level. The positioning of the switch previously had given rise to a number of complaints as, to reach and operate

it, meant quite a stretch for most drivers. Overall, a considerable amount of attention had been given to the minor controls: while there were individual switches on all doors for the window lifts, on the driver's door a four-in-one switch operated all windows. Additionally, a master switch could disengage those at the rear to prevent children from playing with them.

A further change, whilst slight, was made to the car's dimensions for 1974 and was effective from chassis 18269. The wheelbase was lengthened by a fraction over 6mm to 3049.6mm (120.062 inches) and the front track to 1524mm (60 inches) - an increase of 0.5 inch (13mm). The rear track was also increased, from 1466.9mm (57.75 inches) to 1513.8mm (59.6 inches), the modifications being necessary for adoption of larger section radial ply tyres. In addition, again to suit the new tyres - 235/70HR 15s - wheelarches were flared to prevent fouling, so giving rise to the term 'eyebrows'.

The specification of cars intended for other than the home market underwent some changes in 1974: Sundym glass became standard on all models from chassis 18340, with the exception of those destined for West Germany and Australia. Australian cars

were fitted with speedometers calibrated in kilometres per hour instead of miles per hour to comply with local legislation (chassis 18865) but long wheelbase cars destined for the same country did not receive the modification until later. (Long wheelbase cars, incidentally, are discussed in more detail in the next chapter).

Several important modifications were introduced for the 1975 model year and included a change to halogen headlamp bulbs for all left-hand drive cars with the exception of those specified for the USA and Sweden (from chassis 21104); a seat belt warning indicator was fitted to American market standard saloons at chassis 21177 and, for all cars, rear fog lamps became a feature from chassis 22118. Fitted from the same chassis number was a breakerless electronic ignition by Lucas known as OPUS; a vacuum advance distributor was fitted to UK cars from 22572 and for the USA, Japan and Australia from 22600. All North American, Japanese and Australian cars were modified with a lower compression rate of 7.3:1 instead of 8:1 while other cars had the compression ratio altered from 9:1 to 8:1.

October 1975 is significant as it marks the 10th anniversary of the

Silver Shadows were exported all around the world, as this Saudi Arabian-registered car proves. (Courtesy Rolls-Royce Motor Cars Ltd.)

Silver Shadow and T-Series Bentley. When originally designed, the plan was that the car should have a ten-year production run but, in the event, of course, it was to carry on for much longer. Series II cars were not introduced until 1977, continuing until 1980 in saloon form. The Corniche, however, remained in production for another fifteen years after that. There were a number of reasons for extending the life of the Silver Shadow, not least there being no shortage of orders. Financially, the Silver Shadow had been costed on a decade's production but certain factors had not been allowed for. The collapse of Rolls-Royce had delayed planning of a replacement model and the design department was forced to spend long, and sometimes frustrating, periods complying with US Federal safety and emission regulations.

1976 saw a general phasing out of what has unofficially become known as Series I cars in favour of a revamped model, the Series II Silver Shadow and Bentley T2 announced in 1977. Along with the Series II saloons, both coachbuilt and long wheelbase cars were similarly upgraded. Series II designation was warranted because of the extent of redesign; far more than a 'face-lift' entailed. The last of the Series I cars, a coachbuilt model, was built on chassis 26708; the final Series 1 saloon, however, a walnut with beige Silver Shadow, chassis number SRE 26700, was delivered to its owner, Delta Leasing Corp., Stamford, Connecticut, on 25th March 1977. By the time the last of the Series I cars had been built,

production of Silver Shadow saloons numbered 16,717; in contrast, only 1712 Bentley T-Series saloons had been produced, a fraction over 10 per cent of Shadow output. When including the number of Silver Shadow and T-Series coachbuilt cars that were made, the total number of first series vehicles rises to 22,457.

Silver Shadow II and Bentley T2

As previously mentioned, the production life of the first series Silver Shadow and Bentley T models was extended from that originally intended, due mainly to two factors: the regulations emanating from the USA and the aftermath of internal disturbances within Rolls-Royce. If original plans had been allowed to materialise, the second series cars might have entered service as early as

1970, initial trials having been conducted at least a year before on an experimental car. No sooner had Silver Shadow production got underway satisfactorily (it was actually 4 years) when thoughts were directed towards its eventual successor. As with all Rolls-Royces the development period was naturally going to be a prolonged affair but on this occasion it was even more drawn out because of the problems already described.

The main feature proposed for Series II cars was an automatic air-conditioning system and this had first been fitted to a vehicle specially built as chassis 63-B within the factory and registered JTU 63G. As did the majority of experimental vehicles, 63-B wore a disguised radiator shell which was Bentley-inspired. This, incidentally, was the same car which tested the

In 1977 a revised version of the Silver Shadow was introduced. Designated Silver Shadow II, the styling incorporated plastic-faced bumpers and an air dam beneath the radiator (not on American cars, though). Specification also included rack and pinion steering and automatic air-conditioning. (Courtesy Rolls-Royce Motors Ltd.)

The Bentley was similarly modified and classified as T2. In this instance, the location is London, outside the Lobb company, famous for its shoes. (Courtesy Rolls-Royce Motor Cars Ltd.)

Moulton Telegas suspension and, after less than 30,000 miles (48,000km) the vehicle was taken out of service.

John Gaskell recalls the testing of prototype Series II cars and the difficulties experienced in trying to perfect the air-conditioning system. Most of the problems were connected with temperature control and the rate at which the air flowed from the outlets inside the cabin. After several trials with a later experimental vehicle, 64-B, the car was taken for a 25,000 mile (40,000km) endurance exercise which involved spending some time in France. The test was not without its problems as the car not only suffered a serious leak from the cooling system - the problem was so bad that at one time the car had to be driven with a plastic tank on the roof with a hose feeding into the radiator - but damage was caused to the exhaust system when a clamp broke free.

64-B was taken out of service between 1972 and 1973 in order for modifications to be made to the fuel tank, air-conditioning system and engine. Returning to service in the spring of 1973, the car's first test was suspension compliancy as well as the use of HR70 tyres. These tests ultimately resulted in wheelarch flares. In 1974, extensive modifications were carried out which complied with US emission requirements and, later, the car was further modified after having spent some time in the USA. John Gaskell was involved in a mammoth endurance run with the car as soon as it returned from America: 50,000 miles (80,000km) were completed over a pe-

riod of 18 weeks when the car was driven on a prepared route around Cheshire and Staffordshire. John claims this was amongst the most tedious test driving he had ever carried out with the course having to be driven to a tight schedule. At intervals of 5000 miles (8000km) the car was taken off the routine for servicing and tests to check emission control. The car was kept going continuously, day and night, apart from when changing drivers, refuelling and pre-arranged service stops.

It was about this time that the experimental department was trying out another test car which, on this occasion, used the Citroën-designed brake pedal - the infamous 'mushroom-shaped' device fitted to that company's idiosyncratic DS model. An unnamed test driver, having taken over the particular vehicle from John Gaskell, was unused to this system and while reversing felt in vain for the familiar pedal. Unable to locate the 'mushroom', he pushed hard against the accelerator by mistake with disastrous consequences: extensive damage to the car.

A test driver's life is rarely boring. When recounting his days behind the wheel of experimental - and production - Rolls-Royces and Bentleys, John Gaskell told of trials using six-speed gearboxes; by putting intermediate

ratios into a conventional box it was possible to step-up the gear arrangement to six speeds. The system, it seems, was too complicated and was abandoned. Before the days of motorway speed limits, a favourite route for test drivers was from Crewe to the junction of the M1 at the Blue Boar Services near Crich, where the cars were filled with fuel. It was then a case of thundering down to the southern end of the motorway which, at that time, was near St. Albans. The cars then turned round and headed north, back to the Blue Boar Services where they were again refuelled. The best time recorded for the round trip of 120 miles (192km) was an amazing 63 minutes but the goal of completing it on the hour was never achieved. This was before the days of self-service filling stations and the attendant at the Blue Boar always refused to believe the journey possible. This, of course, was understandable considering the average speed necessary to do it would have been a little over 116mph (186km/h).

In 1977, John Gaskell was involved in evaluating the last experimental Silver Shadow, a dark blue car, 66-B. The car, which had been running by early 1974, was fitted with a 7269cc engine and was used to test new designs of tyres currently being developed by Dunlop and Avon. Trials using Dunlop's Denovo tyres were disappointing with excessive road noise and a loss in ride comfort. Later, the car was used with steel-braced radial tyres to test the rack and pinion steering which Rolls-Royce intended using. On one occasion, while trials were being carried out in Italy, John was able to attain a maximum speed of 127mph (203.2km/h). The car remained with the experimental department until 1981 and was used to assess the reliability of digital instrumentation and effectiveness of the air-conditioning system. When it had fulfilled its duties, 66-B was dismantled and crushed, having completed a total of 166,000 miles (265,600km).

The Series II models were launched to the press in February 1977 but not announced to the public until March. Chassis numbers commenced at 30001 for Rolls-Royce-badged cars, and 30046 for the Bentley version. A cursory glance might not have revealed a great deal of difference between the first and second series cars but, on closer examination, several important modifications were evident. Most apparent was the adoption of plastic-faced bumpers, complete with polyurethane sides, which looked similar to those specified for the American market cars. It has to be emphasised that the true 'energy absorbing' bumpers were those fitted to US specification models, all the other markets had solidly mounted bumpers. The American type, therefore, protruded two inches (50.8mm) further out than normal to allow for retraction. Consequently the fairings above the bumpers varied in width.

Fitted beneath the bumper, an air spoiler - or dam - was designed to improve stability and road-holding. This feature, along with the headlamp wash and wipe system later introduced for the home market cars, did not apply to US specification models. The design of the radiator shell was also changed so that it became marginally deeper to correspond with that of the Corniche. Due to the design of the bumpers, which were greater in section than before, the grilles beneath the headlamps were deleted, matching the American export cars. From the rear, the cars could very easily be identified by their Series II badging but less obvious changes were the provision of discreet fog lamps mounted beneath the front bumper, side repeater indicators at the rear of the car and redesigned door handles which incorporated more deeply recessed buttons. This latter modification was made to prevent the buttons being accidentally depressed should the car roll over in an accident. It is important to stress that whilst some cosmetic restyling had taken place, there were no changes to the Pressed Steel body panels, which remained exactly the same as on the original cars. At one time, however, changes to the Silver Shadow's styling for Series II cars had been contemplated. An attempt at what was considered a more modern appearance had been made when an experimental vehicle was substantially restyled from the D-post rearwards. The car was given a lower wing line to present a sleeker look, the D-post itself was elongated and the boot lid allowed

to sweep downwards. The tail, instead of the usual upright stance, was angled to almost 45 degrees and looked neat, especially with the early Series bumper. Although the styling was quite attractive, the project was nevertheless abandoned.

It was under the bonnet and inside the cars that most of the differences existed. A new facia with revised instrumentation gave the cabin a completely fresh look and the wholly new split-level air-conditioning system, which had taken an enormous amount of development, provided extra comfort by supplying different temperatures within the interior. The facia, however, had already been a feature of the Corniche since June 1975, and the air-conditioning was derived from that fitted originally to the Camargue but subsequently improved and updated. It should be added that the Silver Shadow was not unique in having a good quality air-conditioning system as many American motorists took it for granted that their cars would be so equipped. What was different about Rolls-Royce's system, though, was that the occupants did not have to have the same temperature throughout the cabin.

The facia on Series II cars received less padding than that of the first series which had to meet US Federal safety regulations. With an expanse of walnut the facia represented something of a return in design to that fitted to the cars from 1965, before implementation of the Federal Safety Standards, but styling and positioning of the instruments was much more selective.

Serious styling changes for a Series II car were, at one time, considered. A full-size model is shown here: note the plastic-faced overriders, redesigned rear quarter panel and lower boot line. (Courtesy John Blatchley)

Viewed from the rear, the styling modifications are quite startling. The project was eventually abandoned. (Courtesy John Blatchley)

Whereas the safety regulations had meant less space for instrumentation - some were deleted in favour of warning lamps to conserve space - a welcome return was made to earlier styling ideas, especially those of pre-T Series Bentleys. Modern in its appearance, the Series II facia adopted a raised cowling ahead of the steering wheel, which itself was reduced in size from 16 inches (407mm) to 15 inches (381mm). Two large dials ahead of the driver housed an electronic speedometer, which replaced a cable-operated type, together with odometers and sepa-

This is an interesting picture as it shows not only a Standard Saloon on the left, but also the ill-fated Series II prototype next to it. On the far right is a Silver Cloud III, but look closely at the American car; it is a front wheel drive Oldsmobile Toronado which Rolls-Royce was evaluating. (Courtesy John Blatchley)

79

Late Shadow IIs were fitted with headlamp wiper systems. The other cars in the picture are a P1, first-series Silver Shadow, 20-25 and what appears to be a Bentley Corniche or Continental. (Courtesy Rolls-Royce Enthusiasts' Club)

rate gauges for oil pressure, fuel, temperature and an ammeter. To the right of the speedometer a rectangular panel housed no fewer than 10 warning lamps which registered everything from freezing external conditions to brake failure and low water in the screenwash bottle. Next to it was the ignition lock - Yale, of course - and lamp switches which were contained within a circular plate. Positioned in the centre of the dashboard was the clock and an ambient temperature gauge, the main beam indicator being placed between them. On either side of the gauges were the circular upper level air-conditioning vents, which were always referred to as 'bulls-eyes'. The sensor for the ambient temperature gauge was mounted immediately below the Spirit of Ecstasy mascot and was often affected by the heat from the bonnet when the car was standing for any length of time

with its engine idling. Once the car was running normally, however, it operated quite efficiently.

The Spirit of Ecstasy, incidentally, was known within Rolls-Royce circles as The Flying Lady and to those who worked at Rolls-Royce for any length of time, she was referred to as 'Phyllis'. As an aside, Rover's Viking mascot was similarly treated in an affectionate manner and acquired the name of 'George'.

Facia equipment was impressive: ahead of the front passenger seat was a lockable glovebox and, to the side of it, a map reading lamp switch. Beneath the clock, a radio, tape player, seat belt warning light and cigar lighter were all contained within a rectangular housing. To the left of the steering column were switches to control the air-conditioning, aerial lift and hazard warning lights and, to the right, were

the controls for fuel filler cap release, panel rheostat, fog light switch and oil level indicator.

The air-conditioning system fitted as standard to Series II cars was a highly complicated affair and acknowledged as the best of its type, which accounts for its protracted development. The system allowed air to be introduced into the car at two levels and maintained at two different temperatures. Air, as it entered the conditioning unit, was cooled or heated, depending upon the ambient temperature, to 0 degrees C before being heated or cooled further. The air was also dehumidified. so preventing the windows from misting. A device in the air-conditioning system also prevented the lower outlets from emitting cold air onto the occupants' feet before the engine had sufficiently warmed up. Operation of the heated rear window

80

continued on page 97

John Blatchley envisaged the Silver Shadow would stay in production for a number of years and created a shape that would appear as fresh after 10 years as when the car was launched. In saloon form it continued for 15 years and in Convertible guise for 30 years, such was the foresight of Rolls-Royce's chief stylist. (Author's collection)

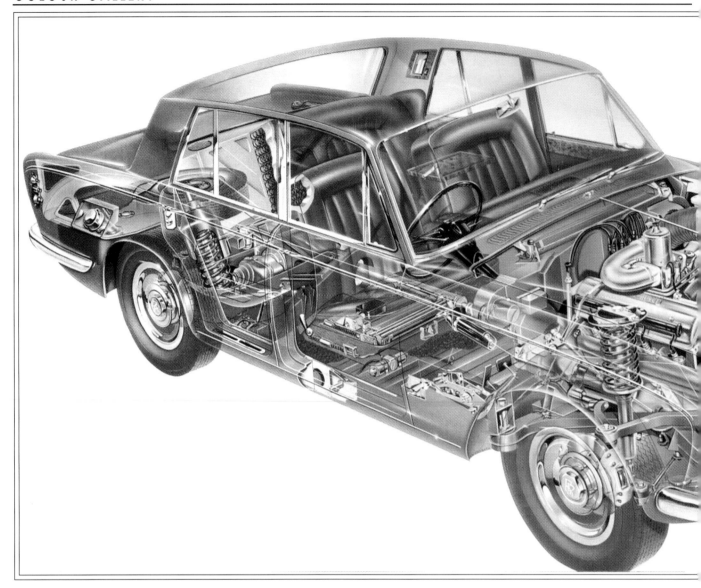

This colour cutaway drawing illustrates some of the Silver Shadow's complicated engineering. (Author's collection)

Just how good a Silver Shadow or T feels can only be properly appreciated by getting behind the wheel ... (Author's collection)

... or reclining in the rear compartment. (Author's collection)

Left: Within six months of the Silver Shadow's launch, the two-door saloon was announced and built in conjunction with Mulliner, Park Ward, in-house coachbuilder to Rolls-Royce. (Author's collection)

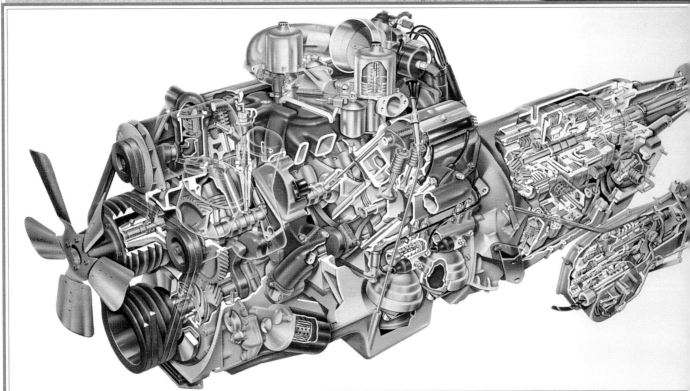

Left: The Convertible had to undergo major chassis reinforcement in order to compensate for torsional stiffness lost when the roof was removed. The sills were strengthened and a cruciform added to the floorpan. (Author's collection)

T-Series Bentley's have always acquired a particular following as some enthusiasts prefer the car's softer styling treatment. (Author's collection)

Left: The 6.230-litre V8 engine as originally fitted to production Shadows and Bentley Ts. (Author's collection)

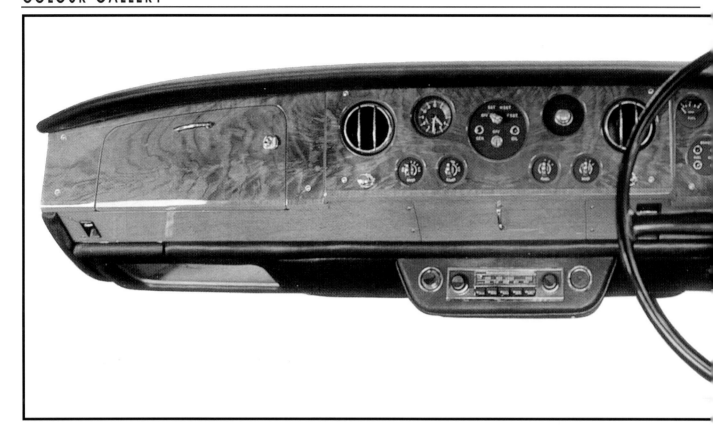

Left: The facia panel and door garnish rails of the Mulliner, Park Ward two-door saloon and drophead coupé are veneered in the finest, specially selected walnut. (Author's collection).

The coachbuilding firm of James Young also produced a Silver Shadow and Bentley two-door saloon. The car illustrated here is the Bentley version of which only 15 were built. (Courtesy Rolls-Royce Enthusiasts' Club)

Left: Early Silver Shadows and Bentley Ts may have been around for up to 30 years; even the youngest cars are at least 16 years old which calls for care when considering a purchase. (Courtesy Martin Bourne)

Beware of cars which have been involved in an accident. Such damage is expensive to repair. (Courtesy Michael Hibberd)

Late Silver Shadows, like this Series II model, can make a fine investment. Such cars may be offered for sale through Rolls-Royce dealerships but are more often sold through independent specialists, at auction or privately. (Author's collection)

To improve ride and handling, Rolls-Royce and Harvey Bailey Engineering each produce suspension kits. Andrew Morris, who owns this beautiful T2, has fitted a handling kit to his car. The performance of the car has been transformed. (Courtesy Andrew Morris)

Right, top: It's no wonder that the Corniche models - whether Rolls-Royce or Bentley - are regarded as possibly the most desirable of all the Silver Shadows. The ageless styling is a tribute to John Blatchley and Bill Allen. (Courtesy Rolls-Rolls Motor Cars Ltd.)

Right, bottom: What better way to enjoy a summer's day than to lower the folding top of a Corniche and glide into ecstasy ... (Courtesy Rolls-Royce Motor Cars Ltd.)

During 1992 an Anniversary Edition of 25 Corniches were produced. Equipped with such features as a vanity set in the driver's door and cocktail requisites in the passenger's door, further finery included special wheeltrims and a plaque set within the glovebox lid. The car illustrated is to American specification, identified as such by the side marker lamps. (Courtesy Rolls-Royce Motor Cars Ltd.)

Every effort was made to ensure that each car incorporated the best and most effective developments in modern technology. (Courtesy Rolls-Royce Motor Cars Ltd.)

The Rolls-Royce Silver Spirit II. (Courtesy Rolls-Royce Motor Cars Ltd.)

The Rolls-Royce Corniche III. (Courtesy Rolls-Royce Motor Cars Ltd.)

The Rolls-Royce Silver Spur II. (Courtesy Rolls-Royce Motor Cars Ltd.)

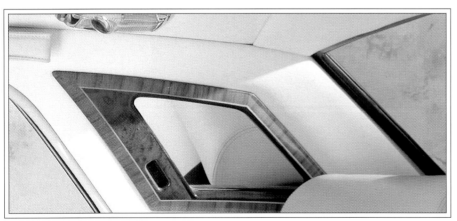

Highest quality and attention to detail is part of every Rolls-Royce. (Courtesy Rolls-Royce Motor Cars Ltd.)

Hide used in the Corniche's interior goes through a long-established process of selection, tanning, dyeing, cutting and stitching. (Courtesy Rolls-Royce Motor Cars Ltd.)

The highest quality walnut veneer is used in the interior. (Courtesy Rolls-Royce Motor Cars Ltd.)

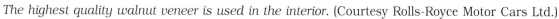

The US specification Corniche II has an air bag and front and rear seat belts. (Courtesy Rolls-Royce Motor Cars Ltd.)

A feature of the Silver Shadow II and T2 was automatic air conditioning. Once set to the required temperature, it needed no alteration whatever the external conditions. The car pictured in this press photograph is a late model, identified as such by its headlamp wipers. (Courtesy Rolls-Royce Enthusiasts' Club)

was also fully automatic and a switch to turn it on and off was not necessary. Describing how the air-conditioning system worked, John Hollings explained the objective was to make it as unobtrusive as possible. Particular care was taken in its design and the intention was to allow gradual changes in fan speed and air temperature. When the interior of the car was either too hot or too cold, the air-conditioning fan operated at maximum speed, reducing gradually when the ideal temperature had been reached. When the engine was started in cold weather, the fan switched itself off, closing at the same time the lower air quantity flap to prevent air circulating at foot level. As soon as the coolant reached 44 degrees C, the flap opened. In hot weather, the fan would start immediately to provide cool air.

During development work a problem was discovered when an already warm engine was started in cold weather. Warm, moist air trapped within the evaporator unit and the trunking surrounding it was blown directly onto the windows where it condensed within seconds and completely obscured vision. To prevent such a situation from re-occurring, fan operation was delayed for 12 seconds, which allowed enough time for removal of the moisture from the air entering the system.

Unseen, but much appreciated by those who drove Series II cars, was the Burman rack and pinion steering, which made the car much lighter and more positive to handle, even with power assistance. As well as the smaller steering wheel, lock-to-lock was reduced to only 3.5 turns. This was the

first time rack and pinion steering had been used on a Rolls-Royce car and it was used in association with a redesigned front suspension. Having greater swing-axle effect, the geometry of the suspension increased resistance to roll by keeping the wheels more upright when cornering which, apart from improving the car's handling, reduced tyre wear.

With emission controls very much in mind, Series II cars were fitted with two SU HIF7 carburettors. These had 1.87 inch (48mm) chokes which complied with regulations in a number of countries in addition to America. The design of the carburettors, which were tuned for economy, not only very efficiently controlled mixture levels but, adversely, had an affect upon engine output. The situation was redressed, however, by the fitment of a twin ex-

continued from page 80

haust system, stainless steel, of course, which imposed less back pressure. In an attempt to conserve fuel, the engine-driven cooling fan was augmented by an electric fan. An added advantage - the point was made by motoring journalists at the time - was that it made the car a little quieter. Gone, it seemed, was the era when all that could be heard while travelling at 60mph (96km/h) was the clock ticking!

The American state of California was so obsessed with emission controls that even Rolls-Royce's efforts to reduce pollution were not good enough. For Californian cars only, therefore, carburettored engines were replaced by those fitted with Bosch K-Jetronic fuel injection which ensured the metering of precise levels of fuel. Eventually this system was widely used by other manufacturers.

The Bentley, now designated T2, received the same modifications as the Silver Shadow. In essence, though, any distinction between the two cars was virtually imperceptible: the engine rocker covers were marked Rolls-Royce, as was the maker's plate showing the chassis number. Even the double R emblem appeared on the facia instrumentation. On introduction of the new models, both the Silver Shadow II and the T2 were identically priced which effectively removed any distinction between the cars. Obviously the prestige associated with the radiator and badging was the deciding factor but,

oddly -and thankfully - there still existed a demand for Bentleys - but only just.

Long wheelbase cars continued in production with the Bentley being known simply as the T2 long wheelbase. The Rolls-Royce badged car, however, was renamed Silver Wraith II, evoking the immediate post-Second World War period of the company's history. Described in more detail in the next chapter, these cars benefited by having a further 4 inches (102mm) added to the wheelbase.

Series II cars went on sale at £22, 809, an increase of £3147 on the price of the car it replaced. It can be argued that the increase was due to an upgrading of specification, especially the superb air-conditioning. Prices of Silver Shadows and Bentley T had consistently increased from the original £6556 and £6496 respectively; by 1967 the price had risen to £6670 for the Rolls-Royce and £9272 in 1970. The basic price in 1973 was £10,403, which did not include the tape player, and three years later, in 1976, it had jumped to £17,898.

For those motorists who were not convinced that the Silver Shadow or its Bentley sister car had no rival, there was a choice of exotic cars. Mercedes' 450SEL, with its 6.834 litre engine and at a fraction more in price at £22, 999, was certainly a quicker car with a 131mph (210kph) top speed. The Vanden Plas Daimler, Jaguar's upmarket flagship, was nearly half the

price but what it gained in performance (136mph/ 218kph) was lost in craftsmanship and finish. Capable of 140mph (224kph) and oozing with sporting traditions of wood and leather was the Bristol 412, and at a shade under £20,000 at that. Aston Martin's V8 saloon was astonishingly fast, over 160mph (256kph) was possible and, at the other end of the scale, Cadillac's Seville was superbly comfortable when driven in a straight line and on smooth road surfaces. The Silver Shadow II was expensive and handling and performance, although dramatically improved, could not be considered sporting. It was, however, totally different to any other car in terms of thoroughness in design, build and finish. In a word, it was quality.

The refinement embodied by the Series II Silver Shadow and Bentley was as expected. The owner was cosseted in the best quality Connolly hide and surrounded by expertly chosen burr walnut veneer. Handling - far different to that of earlier cars - was more positive and the car rolled a lot less than before. This was helped by the provision of a slimmer rear anti-roll bar which compensated for the lowering of the front roll centre, a modification which made the car much more pleasurable to drive. These revisions had other advantages: tyre wear was considerably reduced, an important factor when considering the cost of replacement. Although not obtrusive, the air dam did have a noticeable stabilising effect, especially in crosswinds. The development department paid a lot of attention to the results of tests

Successor to the Silver Shadow II and Bentley T2 was the Silver Spirit. Design work was started in 1969 and the project was known as SZ. (Courtesy Rolls-Royce Enthusiasts' Club)

using the wind tunnel and its efforts had been rewarded.

Maximum speed of the Silver Shadow II and T2 was claimed as 120mph (193km/h) and for those owners who were interested, the engine would be working at something like 4600rpm. Overall fuel consumption in the region of 13-14mpg (20Lt/100km) could be expected, hence the need for a 23.5 gallon (107 litre) fuel tank which would allow a little over 300 miles (480km) between visits to a fuel pump.

During its production life, specification changes were made to the Series II car, but not nearly as many as for the first series which amounted to something like 2000, including those which had seemed almost insignificant. The last major styling modification occurred at chassis 34573 when the headlamp wash-wipe system, initially installed on Swedish export cars, was fitted to all vehicles except those bound for the USA. Not so much a blade, the headlamp wiper was a nylon brush which was far more successful in sweeping the convex shape of the headlamp lenses.

Mechanical specification changes, too, were kept to a minimum and the final modification of any note was the introduction, at chassis 40194, of

Bosch fuel injection for Californian cars.

Prices continued to rise and in 1979 the cost of a Silver Shadow II and Bentley T2 was £36,652, more expensive than most cars considered anywhere near comparable. BMW's 733i was well under half the price and by 1979 the Mercedes 450SEL 6.9 attracted a price tag of just under £30,500. In the supercar league, the Rolls-Royce and Bentley were the most sedate and the Daimler Vanden Plas 5.3 and Ferrari 4001, two high performers capable of virtually 150mph (240kph).

Some revisions (to the suspension, for example), were made to the Silver Shadow range of cars, but were not applied to the standard saloons.

Styling of the Silver Spirit facia was not unlike that of the Silver Shadow II and T2. This picture shows the facia of a Silver Spirit II. (Courtesy Rolls-Royce Enthusiasts' Club)

Instead, modifications such as the use of gas suspension, which supplemented the coil springs, appeared on the Corniche and Camargue and were effective on the Silver Shadow's and Bentley's replacements, the Silver Spirit, Silver Spur and Bentley Mulsanne. The Series II saloons, although very successful in their own right, were designed in some respect as a stopgap, while the new generation models were being perfected. During the Silver Spirit's gestation period there is no doubt that the Silver Shadow was used as a development vehicle and John Hollings did not hide this fact when he spoke of Rolls-Royce's engineering role during the seventies: *"In accordance with our policy of introducing changes first on the coachbuilt cars, the new rear suspension was introduced initially on the Camargue and the Corniche in 1979 and on the Silver Spirit when it was introduced at the end of 1980. The new suspension was never in fact fitted to the Silver Shadow, although this car was of course used for much of the development work."*

Work started on the Silver Spirit as early as 1969. In view of the huge outlay involved in the development of the Silver Shadow it was evident from the outset that the new car would use much of the technology of its forerunner. The Silver Shadow's underframe was modified at the rear to take the new rear suspension and widened in the engine compartment in order to make that for the SZ range of cars. The body styling of the Silver Spirit and its ensuing Bentley model was the work of Fritz Feller and although it enjoyed an

individuality of its own, the underlying identity with its predecessor nevertheless remains. This is most apparent when studying the Silver Spirit's facia which, on first glance looks identical to that fitted to the Silver Shadow II. The only real difference between the two is the digital display, consisting of outside temperature, clock and stopwatch in a central panel above the radio housing, and an extra 1.25 inches (32mm) in width at each end. Such was the increase in interior width over the Shadow.

In detail, the Silver Spirit was almost 3 inches (76mm) longer, 2.3 inches (62mm) wider and 1.25 inches (33mm) lower than the car it replaced. It had more window area - about 30 per cent - and a slightly longer wheelbase due to the suspension design. There was a fraction more room inside the cabin, 4 inches (102mm) wider and hardly any extra length, but the engine and transmission were all as before. The frontal styling was more modern and the radiator grille wider which helped make the car appear much larger than the Silver Shadow. Although the Silver Spirit looked as if it might be faster, its top speed was, in fact, a fraction less at 119mph (191.5kp/h).

For Bentley enthusiasts who had mourned the car's decline, there was the excitement of a noticeable resurgence in the marque. Sporting traditions were revived with the appearance of the Mulsanne Turbo, a fabulous machine with an output confirmed at a shade under 300bhp and top speed of 135mph (217.2kph), and the even mightier 328bhp, 146mph (235kph)

Turbo R. The Bentley Eight, with its wire mesh radiator grille, evoked memories of the marque's competitive and vintage era and was designed around the Mulsanne's specification but with simplified trim. The model was introduced in 1984 to redress the effect of a recession which had badly affected the market for luxury cars in the early eighties. Sales of Rolls-Royce and Bentley cars had plummeted: only 1551 cars were produced in 1983, less than half the output of 1981 (3165). The Continental R two-door coupé announced in 1992 was another model to revive a famous name and was the first Bentley in forty years not to have a parallel Rolls-Royce-badged sister car. In 1995, another newcomer, the Bentley Azure, a convertible version of the Continental R, made an appearance. For some enthusiasts of the winged B though, only the Java concept car, unveiled in 1994, sets the standard upon the future.

The Silver Shadow and T-Series Bentley saloons were built until the autumn of 1980, so ending almost 15 years of production. Since introduction, 8425 Silver Shadow IIs left the factory, along with 2135 Silver Wraith IIs and 568 T2s, which included 10 long wheelbase cars. The price of the saloons had reached £41,959 and that of the Silver Wraith £49,447. Silver Wraiths supplied with a division between the front and rear compartments

Draped with the Union Jack, the last Silver Wraith II bodyshell has been delivered to Crewe from Pressed Steel. The car is still in its Body in White stage. (Courtesy Rolls-Royce Motor Cars Ltd.)

were priced at £52,138. The Camargue went out of production in 1986 but the Corniche remained until 1995 to continue the concept of what has, so far, been recognised as the most successful car in the history of Rolls-Royce.

It was a fitting epitaph, therefore, when, thirty years after launch in the autumn of 1995, over 300 Silver Shadows and T-type Bentleys gathered together in the grounds of Blenheim Palace, one of Britain's finest stately homes, to celebrate this auspicious occasion.

The last Silver Shadow car, a Silver Wraith II, under production; an auspicious occasion and the team employed in building it are recorded on film for posterity. (Courtesy Rolls-Royce Motor Cars Ltd.)

IV

COACHBUILT MODELS AND VARIANTS

By the time Rolls-Royce monocoque cars were being developed, the age of the bespoke carriage, painstakingly built by experienced craftsmen, was almost at an end. Almost, but not quite, because there remained a very carry on trading. Those that did continue business, however, were affected by the recession of the thirties which plunged the industry deeper into decline, causing many established firms to disappear.

Rolls-Royce and Bentley coachbuilt cars are amongst the most elegant bespoke motor vehicles in the world. This 1956 Hooper-bodied Silver Cloud (chassis number SYS 18) is owned by Harold Kay. (Author's collection)

limited market for cars such as the Flying Spur, special-bodied Silver Clouds and exclusive Phantoms to which the traditionalist coachbuilder could apply his skills.

Before the introduction of unitary construction methods, all cars, of course, were built with separate chassis and there was no shortage of first class coachbuilders. As mass-production techniques accounted for more and more car output, so the demand for specialist coachwork declined and many famous firms found it difficult to

War followed recession and, with it, a disruption of the motor industry although vehicles, of course, were needed for the armed forces. When the industry resumed after hostilities it was a matter of earning precious foreign currency, which meant car building for export took priority. If Rolls-Royce was to succeed, it had to adopt rationalisation if only because none of the remaining coachbuilders had the capacity to keep up with demand. With the decision to buy-in complete standard steel bodies, those coachbuilders

The elegance of this coachbuilt Bentley S3 is quite delightful. (Author's collection)

who had so far survived were further demoralised.

Throughout the company's history, all of the most reputable coachbuilders had produced specialist bodies for Rolls-Royce and Bentley: Barker, Freestone and Webb, Gurney Nutting, Hooper, Arthur Mulliner, H.J. Mulliner, Park Ward and James Young were all represented. Barker, after going into liquidation in 1938, was taken over by Hooper, which in turn became part of the Daimler company during the late thirties. In 1959, for the last time, Hooper exhibited, at the London Motor Show, coachwork intended for "royal and distinguished patrons"; the name has survived though and has been responsible for specialised models based upon the Bentley Mulsanne Turbo and Bentley Turbo R. Freestone and Webb was amalgamated into the H.R. Owen group of companies in 1955; Gurney Nutting was acquired by Jack Barclay in 1945 and Arthur Mulliner, ceasing to build private cars in 1939, was absorbed by Henley's Limited. Park Ward was purchased by Rolls-Royce in 1939 and, after the war, carried out specialist and experimental work for its parent company. In 1959, H.J. Mulliner and Company was taken over by Rolls-Royce and, two years later, H.J. Mulliner, Park Ward Ltd. was formed, retaining premises at Willesden, in north London, and Chiswick, west London, respectively. James Young also survived the war years, having been acquired by Jack Barclay in 1937, and resumed production of specialist coachwork for Rolls-Royce and Bentley cars.

At the time of the T-Series and Silver Shadow launch, the only traditional coachbuilders associated with Rolls-Royce and Bentley were James Young and H.J. Mulliner, Park Ward. James Young Ltd. had been in business for 102 years when the Silver Shadow appeared and was more famous for its bodies designed for the Silver Cloud and S3 chassis, along with some superb coachwork for the Phantom V. The latter, incidentally, is now considered amongst the most desirable and sought-after of all postwar Rolls-Royces. H.J. Mulliner, Park Ward had become the 'in-house' coachbuilder to Rolls-Royce: Park Ward which, before 1952, enjoyed a distinctiveness all its own, took on more of an identity with the standard steel products under the direction of John Blatchley; H.J. Mulliner - responsible for over half the total production of the 17 exclusive Phantom IVs - successfully managed the transition from using seasoned hardwood to metal for the framework of its bodies.

The decline in traditional coachbuilding is one of the reasons why Rolls-Royce adopted unitary construction for the Silver Shadow and T-Series models. Nevertheless, the exclusive market for coachbuilt motor cars still existed, albeit to a limited extent, and Rolls-Royce, naturally, felt compelled to satisfy this demand. Coachbuilding techniques relating to cars with monocoque construction were well understood by the time Rolls-Royce developed its two-door saloons and convertibles, such specialists as Pininfarina, Bertone and Chapron having devised some intriguing designs around Fiat, Lancia and Citroën models. Before discussing the two-door variants, the forerunners to the Corniche and the Camargue, an important derivative of the four-door saloons must not be overlooked.

Long wheelbase saloons

The long wheelbase versions of the Silver Cloud and S-Series Bentley were always sought-after. Out of the 14,850 cars produced, just 797 (a little over 5 per cent) were constructed on the long chassis and it is surprising, therefore, that a long wheelbase Silver Shadow or T-Series Bentley was initially omitted from the catalogue. A long wheelbase Silver Shadow option was, however, intended, and it will be remembered that initial ideas for these cars were based upon designs with a longer wheelbase than the definitive prototypes.

The pressure of getting the saloon cars ready for the Paris Motor Show launch, as well as developing the two-door models, meant that any project for an alternative long wheelbase variant was delayed longer than had been anticipated. It was not, therefore, until the spring of 1969, over three years after the standard saloon had made its appearance, that production began on a 10ft, 3.5in (3137mm) wheelbase car using chassis number 6599. A pilot batch of 10 cars, incidentally, was made during 1966/7 and included a car specially built for HRH Princess Margaret. This particular car is discussed later in the chapter.

Long wheelbase saloons were es-

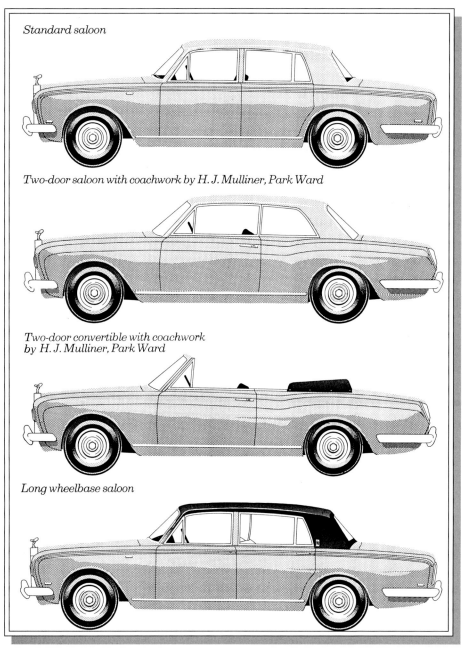

Standard saloon

Two-door saloon with coachwork by H. J. Mulliner, Park Ward

Two-door convertible with coachwork by H. J. Mulliner, Park Ward

Long wheelbase saloon

Sadly, coachbuilding went into decline and, in order for Rolls-Royce to survive, radical (for Rolls-Royce) production techniques had to be adopted. Nevertheless, there was still a demand for the bespoke carriage, hence the introduction of the two-door models as well as a long wheelbase car, constructed by Mulliner, Park Ward. This illustration, from a sales brochure, depicts the Rolls-Royce range in the late sixties. (Author's collection)

Mulliner, Park Ward, at Willesden, where the bodyshells were received direct from Pressed Steel at Cowley. Once the coachwork had been completed, the bare metal bodies (body-in-white) were dispatched to Crewe for finishing in the usual manner. All this was time consuming of course, and long wheelbase cars took on average 4-5 weeks longer to prepare than standard saloons. As to be expected, the cars were fully equipped and those fitted with a division - which had an electrically operated window lift - had the benefit of two air-conditioning systems, one for each compartment, the refrigeration unit for the rear passengers being housed in the boot. Almost unthinkably, the cars with divisions lacked picnic tables due to the fitment of the glass screen, and rear vanity mirrors, usually built into the rear quarter panels but now replaced by discreet air-conditioning vents.

Although available from the first half of 1969, long wheelbase cars were not specified for the home market until later in the year, with early production destined for America. It is not surprising, therefore, that all of these cars were finished to US Federal safety standards and incorporated a high degree of facia padding. All long wheelbase cars received the GM400 gearbox and, apart from a few early examples, were fitted with automatic height-control suspension to the rear wheels only.

Notwithstanding the fact that production of long wheelbase cars did not get underway until the standard saloon had been on the market for over three years, a pilot batch of 10 cars had

sentially standard saloons with 4 inches (102mm) added to the wheelbase just aft of the centre pillar, the B-C post, which provided a noticeable increase in space in the rear compartment. Once the floorpan and sills were cut and the new section welded into position, the body received lengthened rear doors and a new, longer roof panel. Always fitted with an Everflex vinyl-covered roof, the cars, apart from the earliest examples, also had a slightly smaller rear screen which provided a measure of extra privacy. An optional feature of long wheelbase cars was the provision of a division between the front and rear compartments which, when fitted, took up the entire extra floorpan length.

Instead of being carried out at Crewe, modifications to the bodywork were completed at the workshops of

Long wheelbase cars had an extra 4 inches (102mm) added to the floorpan. Bodyshells were delivered to Mulliner, Park Ward from Pressed Steel, where the was conversion carried out, before the car went to Crewe for finishing. The car shown here is actually a Silver Wraith II. (Author's collection)

been made available in 1967. Delivered on 19th July of that year through Kenning Car Mart Ltd., the car, which had been specially commissioned by HRH Princess Margaret and Lord Snowdon, was built on chassis LRH 2542 and incorporated several special features, some of which were exclusive to royal requirements. Very much involved in the preparation of this particular car was chief stylist, John Blatchley, who also had responsibility for all styling matters at Mulliner, Park Ward. Together with Roger Cra'ster who, as export manager at Rolls-Royce normally conducted all negotiations between the company and the royal family, John paid two visits to Princess Margaret to discuss her specification requirements and suggestions. On the first of the two visits John recalls that Princess Margaret and Lord Snowdon knew a great deal about the car as Dr Llewellyn-Smith had briefed them the evening before during a dinner engagement.

Features included in Princess Margaret's car were a height and forward-adjustable rear seat for ceremonial occasions, extra night illumination and the standard fitment of a special police lamp above the front screen. The usual pennant, raised when the car was used for official functions, was quite disliked by Princess Margaret, and John Blatchley accordingly designed a most beautiful, internally lit 6 inch (152mm) high shield bearing the royal coat of arms, exquisitely hand carved from solid perspex. A further variation was to the repeater direction indicators, which were built into the waist moulding and were similar in design to those used by Fiat. The interior of the car was equipped with special green hide to match the colour of the bodywork, whilst the carpets were a pale grey-green colour. Instead of the usual burr walnut, the facia was made of oiled teak and the door capping rails, normally made from wood, were of leather to match the upholstery. All brightwork was given a satin finish.

Long wheelbase cars were also built as Bentley models, but only in comparatively small numbers: 2780 first series long wheelbase Silver Shadows were produced but only 9 T-Series. When the second series cars were introduced, the long wheelbase version was redesignated Silver Wraith II so reviving a famous model name. Unlike the earlier car of the same name, however, the Silver Wraith II did not have a separate chassis. The Bentley version, however, underwent no such resurrection, commanding simply the title T2 long wheelbase. Series II long wheelbase cars received the same modifications as the saloon and were immediately identified by their plastic-faced bumpers, dedicated badging upon the boot lid, together with the Rolls-Royce

A pilot batch of 10 long wheelbase cars was made between 1966 and 1967 and included this particular vehicle - chassis LRH 2542 - specially built for HRH Princess Margaret. The car had several styling modifications to suit the Princess and was finished in dark green paint with special green hide to match. Carpets were pale grey-green and the facia was made from oiled teak. Note the royal coat of arms which John Blatchley specially designed on request from the Princess. Take a look, too, at the repeater indicators on the waist moulding which are very similar to a component used by Fiat. Note the police lamp above the windscreen. (Courtesy John Blatchley)

The lengthened wheelbase is evident in this publicity photograph of the Silver Wraith II. (Author's collection)

or Bentley emblem neatly positioned on the rear quarter panels, and by the redesigned facias. Like its predecessor, the Silver Wraith II when fitted with a division, lost all advantage of the extra leg room for rear passengers, the centre fitting accounting for the additional 4 inches (102mm) built into the platform. Only 10 Bentley T2 long wheelbase cars were built, contrasting dramatically with 2135 Silver Wraith IIs. Cars for the American market did not, of course, have the air dam beneath the front bumper and safety regulations prevented the fitting of a central division. The reason for this was that in order to install the rear air-conditioning refrigeration unit, the fuel tank would have had to be relocated outside of the car's rear crumple zones.

The Silver Wraith II and its Bentley equivalent were replaced by the Silver Spur in 1980, the last chassis built being 41468. In terms of desirability, the long wheelbase cars have become much sought-after, if only because of the greater amount of rear passenger space they afford. Additionally, Silver Wraith IIs are popular as they not only offer better handling with their rack and pinion steering, but also provide greater passenger comfort due to the provision of automatic air-conditioning.

Conversions

Even Rolls-Royces did not escape the customising techniques applied by some specialist firms in order to 'stretch' a normal car into something much bigger. Such conversions were often carried out for publicity purposes,

usually serving as courtesy transport, but some were completed to satisfy the demands of their owners. Often bizarrely fitted out with television, auxiliary seating and the wares of the customiser's art, these cars must surely be a far cry from what Harry Grylls and John Blatchley had envisaged. A number of cars received specialist treatment, such as the car built on chassis LRA 14837. The car's owner, John L. Hanson, had wanted single round headlamps, reminiscent of the Silver Cloud I and II, instead of the twin lamp system; white side lamps were fitted into the tops of the front wings and fog lamps faired into position beneath the headlamps in place of the grilles. At the rear, enclosed wheelarches gave the car a particularly heavy appearance. At least one estate car version was built as Classic Coachworks of California devised a straightforward design on a long wheelbase car with left-hand drive

using chassis number LRX11443. The same company also featured a pick-up style vehicle, an 'estate wagon' conversion using an early right-hand drive car.

Although the Corniche was never offered as a four-door version, one such cabriolet built on an early floorpan did exist. Registered XLR 176, the car featured rear doors hinged at the rear, so as to be front opening, but it is not known who undertook the conversion or, with any certainty, its present whereabouts.

Following a demand from the British Foreign Office, a number of Silver Shadows were equipped with armour plating to make them bullet-proof. The cars were designed for use in certain countries where the embassies were considered vulnerable. Built to individual specification, this work was generally supervised by Jock Knight at the Crewe factory, whose job it was to prepare the car according to the degree

of protection required. The extra weight of these cars caused great difficulty and often meant boosting the car's suspension and restricting the amount of load the car could carry. Jock remembers demonstrating an armoured car to the SAS (Special Air Service) at Hereford and being somewhat concerned when an officer fired three shots directly at the car at point-blank range. Thankfully, both Jock and the car survived the ordeal! In all, no more than 10 vehicles were converted in this way.

Being a bespoke motor car, a number of vehicles were prepared to owner's specific requirements and it is not possible to list them all. Many

Eighteen months after announcing the two-door saloon, Rolls-Royce launched the Convertible, produced by Mulliner, Park Ward. Both models could be specified as a Bentley if required. (Courtesy Rolls-Royce Motor Cars Ltd.)

enthusiasts, however, would not care to see the Silver Shadow in any other guise than that intended. If it was good enough for Rolls-Royce, then it was good enough for anyone ...

Two-door saloons

Less than six months after the launch of the Silver Shadow and Bentley T, Rolls-Royce unveiled an attractive two-door saloon distinctly different to that

of the standard four-door car. A little over 18 months later - two years after the Silver Shadow's debut - another exclusive two-door model, a convertible, was announced. These cars were forerunners to the Corniche which appeared in 1971 and, together, are now recognised as being amongst the most desirable of all postwar Rolls-Royce and Bentleys. At the time of introduction they were designed to

replace the S3 Bentley Continental which was still being produced in very small numbers.

Even though the monocoque construction of the Silver Shadow and T-Series Bentley made it difficult to produce a true coachbuilt variant, Rolls-Royce nevertheless proposed to offer such a car to its discerning clientele. The coachbuilders that remained in business in the mid-sixties were hardly in a position to spend huge amounts on new production methods and, for James Young, the situation was critical if the company was to continue producing specialist bodies. For Mulliner, Park Ward, there was some hope in as much as the company was already part of Rolls-Royce and enjoyed a great deal of support from the parent company.

It is no surprise, therefore, that an early decision was made that Mulliner, Park Ward carry out the coachbuilding of the two-door models, not only because Mulliner, Park Ward was a subsidiary of Rolls-Royce, but also because it had the capacity to produce the cars, even if it did mean a major upheaval in adapting to the techniques employed in coachwork engineering - instead of coachbuilding - building upon an already fabricated platform constructed by mass production methods.

James Young, however, who was fully experienced in producing bespoke bodies for Rolls-Royce and Bentley chassis, was not in the same position as Mulliner, Park Ward. Its independence from Rolls-Royce placed the company at something of a disadvantage

as far as the monocoque cars were concerned but that did not deter it from trying to adapt to new methods of business. Deciding it should offer a coachbuilt alternative to the Crewe product, the company actually came up with a design before Mulliner, Park Ward. Unlike the Mulliner, Park Ward models, which had their own unique construction, the James Young cars were merely an adaptation of the standard four-door saloon. For the Bromley company there was no alternative other than to buy in complete cars from Crewe and carry out what was, essentially, a series of cosmetic modifications which included repositioning the door pillars and fitting new doors. With traditional coachbuilding methods not compatible with the engineering of the Silver Shadow, it would not have been possible to perform any major structural operations.

At first glance the James Young saloon looked almost identical to the four-door Crewe model, such were the lines of the car. Externally, the differences affected the doors - two instead of four - which were elongated, and the provision of rear quarter windows; the chrome waist trim strip was dispensed with and the door handles were of the coachbuilder's own design. There were subtle differences to the car's interior: the rear quarter window lifts were, naturally, electrically operated and the

seats, together with the wood trim, followed the style that was distinctive to the company.

The first of the James Young cars was built on chassis 1067 and the derivative was offered as a Bentley version as well. Apart from the radiator shell and badging the cars were virtually identical, the Bentley receiving the usual wood trim and facia found on the Crewe model. The cars were considerably more expensive than the standard saloons, commanding a further £1214. Demand for the cars was not huge and, after having built only 50 examples, 15 of which were Bentleys, the project was abandoned in 1967.

Throughout this period when James Young was offering a Silver Shadow conversion the company was also building bodyshells for the Phantom V which remained in production until 1968. Having produced 195 out of the 516 built, James Young went out of business after producing its last Phantom, so ending 105 years of business.

Mulliner, Park Ward cars

The decision to build an 'official' Mulliner, Park Ward two-door version of the Silver Shadow was made during the period 1963/4 and it was to Bill Allen that the styling was entrusted. Bill, of course, was a very experienced stylist, having joined Arthur Mulliner

The task of styling the two-door cars fell to Bill Allen, John Blatchley's deputy. John had, naturally, discussed the general styling arrangements with Bill and, as head of department, was responsible for the ultimate design. As John was away from the department for several weeks there was no alternative but for Bill Allen to get on with the job. (Courtesy Rolls-Royce Enthusiasts' Club)

from Towcester grammar school, and an RR man since 1935. As deputy to John Blatchley, Bill Allen had been responsible for much of the work on the standard saloon as, at that time, John Blatchley was away from the factory for several weeks because of illness. The two-door saloon was very much Bill's car and, after completing initial sketches, he started to create the definitive shape using the wax modelling technique. As head of department, John Blatchley had, naturally, already discussed the two-door derivatives with his deputy and Bill Allen's styling met with his full approval. Responsibility for the ultimate design of the cars, therefore, was John Blatchley's.

It took six weeks to complete the styling and the most noticeable feature was the uplifted waistline above the rear wheelarch. Bill, incidentally, refers to this particular aspect of the car's design as his 'Coke-bottle' feature which he originally introduced (somewhat daringly) to suggest something of a rear wing. At that time, such styling ideas were usually restricted to a few sports cars which, because of their low build, employed sculptured rear wings as in the case of the E-Type Jaguar, Big Healey and AC Cobra. This sportscar styling feature was based in tradition and was really nothing more than a hangover from the running board

era. Later, chrome embellishments which suggested a wing line became popular on some saloons. The styling theme of the two-door Silver Shadow, however, successfully imitated the graceful lines of the Bentley S-Series Continentals, especially those offered by Park Ward or H.J. Mulliner with their vestigial rear wings.

A very modest person, Bill Allen claims the styling he devised for the car was very simple with a prominent moulding from front to back on a surface shaped slightly to suggest a rear wing in both plan and elevation views. The moulding, he considered, also gave an impression of speed when viewed from the side, which was quite appropriate for a two-door design. His styling of the car was, therefore, very pertinent considering the model was intended as a direct replacement for the S3 Continental. Strangely, the two-door model, even though intended as the flagship of the Silver Shadow range, did not carry on the Continental nomenclature; for this customers would have to wait for 1971 and the Corniche.

In the meantime, the two-door saloon was accompanied, in 1967, by a drophead coupé version, whose styling was, again, devised by Bill Allen. Everything below the waistline of the

Bill Allen, pictured in February 1996 alongside the last Corniche built. (Courtesy Rolls-Royce Motor Cars Ltd.)

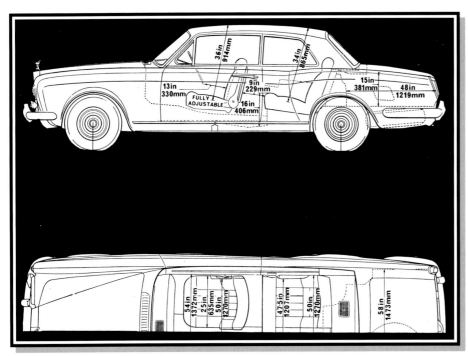

This brochure illustration clearly shows the drophead's internal measurements. Like the four-door saloon, the Mulliner, Park Ward cars were equipped with electrically-adjustable seats. (Courtesy Rolls-Royce Motor Cars Ltd.)

car resembled the two-door saloon version; above it, the hood was designed and built by Mulliner, Park Ward. Both two-door Silver Shadows, as a Rolls-Royce or Bentley, were completely distinctive in appearance and, although sharing a common identity with the standard saloons, nevertheless incorporated a number of subtle differences.

Auxiliary driving lamps were fitted above the front bumper and repeater direction indicators - also a feature of the saloon cars - were positioned further forwards along the front wings, ahead and above the wheelarches. Overtaking mirrors, normally positioned at the base of the A-post, were placed on top of the front wings, while the door handles were of the style unique to the Mulliner, Park Ward cars. The tail layout was enhanced by the rear wing profile, which continued the graceful sweep of the crown line down to a lower boot lid and was much sleeker in appearance than that of the four-door models. As with the four-door saloons, the boot lid, doors and bonnet were all formed from aluminium alloy. The power-operated hood on the Convertible, which was finished in Everflex with a cloth lining, was superbly constructed and as a

safety measure could only be raised or lowered when the car was stationary with the handbrake applied and the gear selector in neutral.

The passenger compartment also featured a number of differences. The front seats - electrically adjustable as well as having a manually operable backrest to allow easy access to the rear cabin - were unique to Mulliner, Park Ward design; the back seat, narrower in width than on the standard saloon, suggested a 2+2 rather than a full four-seater. There was no sacrifice in comfort: English hide upholstery, deep pile carpets and the luxury of burr walnut for the facia and waist rails were features, while the capping rail was padded in black leather. The windows, including rear quarterlights, were electrically-operated, as was the radio aerial, and specification included

The facia of the Mulliner, Park Ward Convertible.   Courtesy Rolls-Royce Motor Cars Ltd.)

was fitted to right-hand drive versions while the three-speed GM400 unit was specified for left-hand drive models. The constant-height self-levelling suspension was also a feature and, similar to the standard saloons, offered excellent ride comfort. From 1969, as on the cars built at Crewe, the levelling device on the front wheels was deleted, leaving the high pressure hydraulic rams to work on the rear axle only.

Construction of the two-door mod-

cigar lighters in each of the rear armrests and, as expected, air-conditioning as an option.

Unveiled at the Geneva Motor Show in March 1966, the two-door Silver Shadow Saloon went on sale at £8150 (£9849 with tax) with the Bentley version £50 less at £8100 (£9789), an increase of 50 per cent more than the standard saloon. The two-door Convertible, making its debut at the London Motor Show in 1967, attracted a further £400 with prices of £8550 (£10,511 including tax) for the Rolls-Royce and £8500 (£10,450 including tax) for the Bentley. Despite these higher prices (which reflected the high build costs), there was an immediate demand for the two-door models.

Chassis numbers for the two-door Mulliner, Park Ward cars commenced at 1148 and 1149 (Rolls-Royce and Bentley saloons respectively), and 1698 for the Rolls-Royce Convertible. Production of Bentley Convertibles, however, did not get underway until chassis 3049. Mechanically, the Mulliner, Park Ward cars were identical to the Crewe types: the 6.230-litre engines powered the cars smoothly and quietly; the four-speed automatic gearbox

Mulliner, Park Ward cars were certainly luxurious. (Courtesy Rolls-Royce Enthusiasts' Club)

It took anything between 4 1/2 and 6 months to build a two-door car, and sometimes longer if the customer had special requirements. By comparison, the time taken to build a standard saloon was 12 weeks. The car in this photograph is built on chassis 6539. (Courtesy Rolls-Royce Enthusiasts' Club)

els was surprisingly complicated and necessitated the cars being transferred between London and Crewe. In essence, the platforms were built by Pressed Steel and delivered to the Willesden works of H.J. Mulliner, Park Ward, where the bodyshells were built into position before being sent to Crewe to have subframes and mechanical components fitted. Whilst there, the bodyshells received corrosion proofing and primer coats of paint before being returned to Willesden for fitting out and application of the finishing coats of paint. As can be appreciated, this was a costly and time-consuming performance which meant an increased price and delayed delivery.

The time taken to build a Mulliner, Park Ward car could be anything between 4 1/2 and 6 months, depending upon customer requirements, in sharp contrast to the 12 weeks it took to complete a standard Silver Shadow or Bentley T. Firstly, the car would spend at least three weeks at Willesden before being sent to Crewe where it remained for not less than five weeks; once returned to Willesden, several weeks of tests and checks were needed before delivery could be authorised. (The experimental department at Pym's Lane often played an unusual role in this process. Frequently, during the early period of the two-door models, parts availability at Hythe Road was somewhat erratic and, to improve the

situation, test drivers from Crewe using experimental vehicles were asked to deliver items when there was an urgent requirement).

Design and build of the two-door cars was quite different to that of the Crewe saloons and warranted the careful transition from the principles of monocoque construction to specialist coachbuilding techniques. As a result of the cars being built onto the standard, mass-produced platform, much new panelling was necessary. Some of this was fabricated at Willesden by Mulliner, Park Ward but other material had to be bought-in from suppliers such as Dowty Doulton, Airflow Streamline and Albany Jig and Tool. The platform shells received from Pressed Steel needed sill and tunnel areas reinforced and, in the case of the convertible model, the entire floor required strengthening and cross-bracing to compensate for not having a steel roof and the weight of the hood.

Not only at Crewe but also Willesden, continuing development gave rise to a whole series of improvements to the Silver Shadow range of cars and some of these were applied to

All Mulliner, Park Ward cars were equipped with the GM400 gearbox from the start. This picture, of what appears to be a publicity car, shows it's exquisite styling. (Courtesy Rolls-Royce Enthusiasts' Club)

the Mulliner, Park Ward models simultaneously with the standard models. Others, however, in keeping with policies dictated by John Hollings and fellow director, David Plastow, were introduced on the flagship coachbuilt models in advance of the Crewe cars and included the General Motors GM400 gearbox and full air-conditioning, which were applied to the two-

The Corniche was introduced in 1971, soon after Black Thursday, the day Rolls-Royce went into receivership. (Courtesy Rolls-Royce Enthusiasts' Club)

door cars a few months ahead of the standard saloon. Modifications to the car's features in accordance with US Federal safety standards were, however, synchronised between the two factories, as was the introduction of the 6.750-litre engine.

The Corniche

The events of Black Thursday, 4th February 1971, the day Rolls-Royce was plunged into receivership, did not prevent the company from announcing a new model within a couple of weeks. The events leading to the company's difficulties have been discussed, and it

This elegant Corniche was pictured at a Rolls-Royce Enthusiasts' Club meeting. The line-up of cars in the background includes some very interesting models. (Courtesy Rolls-Royce Enthusiasts' Club)

is clear that the receiver, E. Rupert Nicholson, was resolute in as much that not only should car production remain unaffected, but the launch of the Corniche go ahead as planned.

The Corniche was vitally important to Rolls-Royce: the two-door Silver Shadow had, in the view of potential customers, too much of an association with the standard saloon and was in need of an identity of its own. By changing the name and introducing, in the main, a number of modifications, a certain amount of sales resistance could, it was considered, be successfully overcome. In view of the model's name it was appropriate, therefore, that the Corniche launch should take place in the South of France, at Nice, in the heart of the French Riviera. In fact, of course, the Corniche was not completely new and neither was it the first time the name had been used by Rolls-Royce. Motoring journalists also considered the Riviera venue appropriate and the sight that greeted them as they disembarked from their aircraft at Nice airport created something of a stir. Lined up outside the terminal lounge were no fewer than nine Corniches, attracting unprecedented attention: excitement mounted as the convoy of Rolls-Royces and Bentleys set off along the suitably named *Promenade des Anglais*. Journalists at-

tending the launch enjoyed a spectacular event: having driven the

Corniches along the coastal route to the Italian border at Menton, they went

on to Genoa, where the cars were put through their paces on the Turin Autostrada, before returning to Monaco.

At the heart of the Corniche was Bill Allen's superlative design which he had styled in the mid-sixties; now with many technical modifications, the car was given an identity which, in view of the company's difficult circumstances, was instrumental in providing a timely boost to Rolls-Royce's image. Not only did the car's name have romantic connotations, it also rekindled memories of an exotic prewar Bentley of the same title. The previous Corniche, a prototype streamlined sports saloon built by Van Vooren of Paris in 1939, was based upon the Mark V Bentley chassis and never went into production. Postwar, the Corniche name was considered for the sports

The two-tone paintwork on this Corniche is particularly attractive and the rarity of the Bentley model adds to the car's appeal. (Courtesy Rolls-Royce Enthusiasts' Club)

coupé which eventually became known as the R-Type Continental.

Like the two-door Silver Shadow and T models before it, the Corniche was offered in two body styles; a coupé and convertible, both of which were built by Mulliner, Park Ward, and badged as Rolls-Royces and Bentleys. Between 700 and 800 personnel were employed at the Willesden coachworks and it was possible to build up to eight cars at once.

Even if the cars looked similar to their predecessors, there were significant differences. The radiator shells of both marques were deeper by 15 per cent, and the interiors were given a facelift. A redesigned facia - which, apart from being very stylish and incorporating a rev counter - pre-dated the similar type to that eventually fitted to the second series Silver Shadow and T saloons. A new steering wheel, 15 inches (381mm) in diameter, with a wooden rim and leather-covered spokes was also a feature. All in all, the Corniche assumed a distinct sporting attitude which, together with modifications to the ignition timing, valve timing and induction system which increased engine output by 20bhp, resulted in a true 120mph (192kph) luxury grand tourer. Engine modifications couldn't be applied to American export cars due to exhaust emission controls so these cars continued with the usual Silver Shadow unit. First seen on the Corniche, and later on the Crewe models, was the cruise control device which could be fitted at the customer's discretion.

Improved cooling was provided by a fan with a modified drive ratio and brake efficiency was enhanced by ventilated wheel trims which allowed extra cooling of the discs. The addition of the rev counter (graduated up to an optimistic 6000rpm) showed the engine to be easily capable of about 4500rpm; on test runs in the South of France, speeds in excess of the maximum were often attained. There was no mistaking the car's pedigree with its exclusive 'Corniche' badging specially designed by stylist Martin Bourne and based on a typeface he discovered in *Reader's Digest* magazine!

Compared with the models it replaced, the Corniche represented a price increase of somewhere around 10 per cent: the Rolls-Royce in saloon version cost £12,829 and the Bentley £72 less; the Convertible commanded almost a further £600 at £13,410 for the Rolls-Royce and £13,332 for the Bentley.

All coachbuilt cars were built to the Corniche standard as from chassis 9770, a Rolls-Royce saloon version, while Rolls-Royce convertibles started from chassis 9771. Bentley badged cars commenced at 10122, for the convertible, and 10,420 for the saloon. Development and refinement of the models continued and resulted in a number of modifications: ventilated wheel discs, which improved brake cooling, were fitted from chassis 10500 and, importantly, compliant front suspension was applied from approximately November 1971 and chassis 12734. It was not until a couple of months later, at chassis 13485, that other models in the Silver Shadow range were similarly equipped with the new suspension. The braking system was yet again revised from chassis 13861, when the brake pedal was given a lighter feel and, at chassis 13922, export Corniches received modified rear calipers.

From 1972, North American cars received energy absorbing bumpers and a pedal-operated parking brake. American Federal safety standards brought about changes which included shortening of the radiator shell to allow horizontal movement of the bumper assembly, and deletion of the air intake grilles beneath the headlamps. Home market and European cars continued with the traditional bumpers until styling changes dictated their demise. In early 1974 - from chassis 18563 - Corniche cars were delivered with the same 120.062 inch (3049.6mm) wheelbase as the four-door cars, 60 inch (1524mm) front track and 59.5 inch (1514mm) rear track together with flared wheelarches. These modifications were made to suit the provision of 235/70VR x 15 radial ply tyres. Lighting was improved by fitting halogen bulbs to all left-hand drive cars, except those for Sweden and the USA, from February 1975 (chassis 21104).

USA cars, commencing at chassis 21729, were supplied from early 1975 with the new automatic air-conditioning unit, a revised facia and fuel emission controls; cars for other markets, with the exception of those for the UK, however, did not receive these modifications until a little later, from chassis 21998. Home market Corniches had to

wait until later, during 1975, for the same improvement, which was made from chassis 22648.

When the Series II Silver Shadow and Bentley T2 were introduced in 1977, similar styling and design changes affected the Corniche. Modifications during production - which were applied to the Crewe models also - were somewhat fewer in number as development was concentrated on the models that became the Silver Spur and Mulsanne, etc. Chassis numbers for 1977 cars started at 30003 for the Rolls-Royce convertible and 30011 for the saloon; Bentley versions commenced with chassis 31219 for the convertible and 31226 for the saloon.

Post-1977 cars were never given a Series II designation (this arrived later, towards the end of the eighties) but these cars, apart from the American market versions, could be positively identified by frontal appearance which incorporated the air dam. Handling of these cars was improved not only by revised suspension geometry but also Burman rack and pinion steering. Of particular interest was the convertible's dual level automatic air-conditioning, the lower level of which operated with the hood down.

In saloon version, the Rolls-Royce and Bentley Corniche remained in production until 1980 when it was discontinued. By then, the price had risen to over £62,000 and the convertibles, which continued in production, commanded a figure of over £66,000. From 1979, however, an important modification affected the rear suspension but, curiously, the changes were ap-

plied without any notification. The reason for this was simple: the end of Silver Shadow II production was approaching and the Silver Spirit, with some of its technical features quite new, was still on the secret list and waiting to be announced. The rear suspension, new for the Silver Spirit, was actually fitted to the Corniche before the event! Handling of the post-1979 cars was dramatically improved, the result of a reworked suspension system which incorporated gas struts, an increased semi-trailing arm pivot angle and slightly increased track measurements. Adding to this, springing was made slightly softer and the springs themselves a little shorter.

The convertible models survived throughout the eighties even though proposals to alter the concept of the car had been contemplated. A plan to update the styling to that of the embryonic Silver Spur was eventually dropped in favour of the existing traditional shape; however, not only had detailed drawings been devised, but a full-scale mock-up of the vehicle had been built. Plans for a revival of the Corniche saloon were also abandoned in the mid-eighties and, during that time,

ideas for a four-door Corniche built upon an extended wheelbase were explored.

For 1986, the Corniche II was announced, though initially for America only; other markets had to wait until 1988 for the new designation. A number of modifications were incorporated in the new model, including redesigned seats and wheel trims and colour-keyed bumpers and mirrors. The Continental nomenclature was revived for the Bentley version, which was no longer referred to as the Corniche, and revised specifications, generally mirrored those applied to the Rolls-Royce version. Apart from new seats, the Continental featured a redesigned facia with separate round instruments which evoked Bentley traditions; further changes included colour-keying mirrors, bumpers and radiator shutters, while distinctive wheel trims, similar to those on the Turbo R, were available. Corniche and Continental cars for the following year were further updated and equipped with anti-lock brakes (ABS) and Bosch fuel injection.

For the 1990 model year, specification changes justified the Corniche III designation for Rolls-Royce cars;

Bentleys, however, remained as Continentals. In general, changes were similar to those applied to the Silver Spirit II, but did not include, until 1992, the saloon's active ride suspension. As a no-cost option, home market customers could choose to have both a catalytic converter and a 10-speaker radio/cassette with CD player. An electronically aided system, suspension settings automatically adjusted to road surface; this was, in principle, similar to the system used by Citroën for its large XM saloon. The Corniche and Continental interiors were also updated to include inlaid woodwork, revised instrumentation and controls incorporating seat adjustment with a memory function.

Active ride suspension and 4-speed automatic transmission - together with a redesigned convertible top that incorporated a heated rear window - were enough for a further designation change - to Corniche IV - in 1992. Although design changes to the convertible top mechanism and some repositioning of the rear window followed a short time later, the specification of the cars was fully comprehensive. Not only ABS braking, but air bags for both driver and passenger were included within the car's safety equipment; emission controls complied with the most stringent legislation at the time.

To acknowledge the car's 21st birthday in 1992, Rolls-Royce built a special limited edition of the Corniche IV which incorporated such features as 4-speed automatic transmission, redesigned seats and burr elm veneer interior woodwork with oak cross-banding and silver inlays. Each of the 25 cars produced (all of which were finished in Ming Blue paintwork with a cream-coloured leather convertible top and sumptuous upholstery in magnolia) were identified by distinctive anniversary wheel trims and, in addition, a specially engraved silver plaque was fitted to the inside of the glovebox lid. The finishing touch was the provision of a vanity set housed within the driver's door and, in the passenger doors, cocktail requisites. It was no wonder, even priced at £165,271, that the cars very quickly found enthusiastic buyers. Rolls-Royce enthusiasts in California celebrated the Corniche's birthday in spectacular style and staged a convoy of 125 cars through elegant Beverly Hills. The event, which entered the *Guinness Book Of Records* for the longest-ever cavalcade of motor cars of one make on a public highway, was not only a tribute to the car but also acknowledged that a quarter of all Corniche Convertibles produced had been exported to that State alone.

Along with the Corniche, Continental and Continental Turbo, a batch of 25 of the Corniche S was built during 1995. These 25 cars were amongst the last Corniche models built and were constructed at Crewe as the production line had already been transferred to Pyms Lane from Willesden in 1992. The MPW factory, incidentally, was closed in 1994. Special order cars are now individually costed and body panels are acquired from Park Sheet Metal, the company which had supplied material for the Camargue.

The Corniche concept goes far beyond a quarter-century: it is now 30 years since the car was introduced as a Silver Shadow two-door model, and something like 33 years since Bill Allen sketched an outline on his drawing board at Crewe. Little did he think then that those pencil drawings of a car with its 'Coke-bottle' waistline would evolve as one of the world's most evocative and exclusive motor cars. An amusing anecdote to illustrate this very point is the occasion when Bill, who had suddenly become very concerned about the car's styling, especially the uplift in its waistline, confided in Martin Bourne, a fellow stylist and said *"you know, this is never going to sell, it's far too old fashioned!"* There is probably no more fitting tribute to Bill Allen's original design than to quote his own words, which are as modest as usual:

"Production of the two-doors saloon, which I always think was the better looking car of the two, eventually ceased. The name 'Corniche' was first given, officially, to the drophead car in 1971 to mark the inclusion of many engineering improvements introduced simultaneously.

Having retired in 1977, it astonished me to discover that this car had almost reached its 25th anniversary in 1993 and some were still being sold."

Pininfarina and the Camargue
With the Silver Shadow fully established and the Corniche models in production, Rolls-Royce's board of directors began to consider a coachbuilt alternative to the existing Mulliner, Park Ward cars which, while sharing

the Shadow's basic platform, would attract even greater prestige. Seen by the Crewe company as its flagship, the model has suffered from much controversy which, in an odd sense, has assured its historical value. To understand the Camargue, and how the car came about, it is necessary to look back further into history.

Pininfarina had, in postwar years, produced a number of designs for cars on Rolls-Royce and Bentley chassis; in 1948 there had been a very stylish coupé based upon the Mark VI chassis, followed by a more conventional drophead coupé a year later. For 1950, the fastback design of two years earlier was repeated, but with revisions that made the car's overall styling much more acceptable to the discerning owner. Pininfarina had also been responsible for a delectable version of the R-Type Bentley Continental in 1954, but not before H.J. Mulliner had visited the Turin company to see how it specialised in the use of lightweight materials for its products.

When it came to styling Rolls-Royce and Bentley cars based upon monocoque construction, there were virtually no contenders apart from Mulliner, Park Ward and James Young. Pininfarina was, however, instrumental in producing a solitary car based upon the Bentley T-Type and this was purchased by James (later, Lord) Hanson. Displayed at the 1968 London Motor Show, the Pininfarina car was certainly controversial with its rectangular headlamps, fastback styling and unmistakable Italian flair; compared to frontal styling, the car looked

much more elegant when viewed from the rear, the detail of the tail styling with its circular tail lights having a characteristically Latin touch. It has been suggested that this car was intended as a forerunner to a new Bentley Continental but, ultimately, it was seen as a precursor to the Camargue. Never completely appreciated, those that saw the Pininfarina Bentley either loved it completely or hated it intensely.

Pininfarina's offering must have made an impression for, soon afterwards, in 1969, the styling house was approached to produce a two-door coupé of such design and proportion, it would be recognised as the flagship of the Rolls-Royce fleet. In fact, the Camargue, as the concept ultimately evolved into, was revered to such an extent that it became the company's most important model, superseding even the Phantom VI and winning the accolade of being the most expensive car on the British market. Code-named Delta, the project as a whole began to develop at a somewhat turbulent time in Rolls-Royce history: John Hollings had taken over from Harry Grylls and John Blatchley had also retired; Fritz Feller was appointed chief stylist and,

in place of "Doc" Llewellyn-Smith, who had departed the motor division, was Geoffrey Fawn who took over day-to-day running of the company.

There were two reasons why Sergio Pininfarina was commissioned to style the Delta project: firstly, it was considered that Rolls-Royce's own styling team was too busy expanding the Silver Shadow theme and, secondly (which may have been the main factor), Geoffrey Fawn had decided the company would benefit considerably from having a car designed by a prestigious Italian styling house. The news that Delta was being styled outside Rolls-Royce was greeted with little enthusiasm from within the ranks at Crewe. Certainly the stylists who had achieved so much throughout the development period of the Silver Cloud and Silver Shadow felt let down at not being given the chance to offer designs. However, once the first feelings of disappointment had faded, there was a general air of excitement at the prospect of a Rolls-Royce exhibiting some of the panache of a Ferrari Berlinetta. That, of course, was not to be ...

It appears that Pininfarina experienced a number of difficulties regard-

ing the Camargue's styling, especially when it came to incorporating the radiator shell within the car's overall size which, although the same length as the two-door saloon and Corniche, was 100mm (a fraction under 4 inches) wider. As a result, the radiator on the Camargue is both wider and lower than that on the Silver Shadow. From a stylist's point of view, the Camargue's sharp-edged shape, scuttle height and huge radiator shell, emphasised its sheer size. The styling offered a number of novel ideas as far as Rolls-Royce was

concerned and included the adoption of a bonded windscreen in place of the usual type secured by moulded rubber, curved door glasses and quarter lights, and a single piece bumper. The construction of the bumper was the same for all markets but mounting differed for the American market cars which incorporated the regulatory energy absorption. The styling of the headlamps was typical of that seen on many American cars of the early to mid-seventies and it is clear that a similar theme was adopted on some of the

prototype designs for the Silver Spirit. At the waist, a locally dropped waistline was intended to make the windows appear larger than they actually were; Pininfarina's original proposals to finish all metal within this area in black were rejected by Rolls-Royce management - the 'black chrome' intended for the window channels was not considered sufficiently durable.

Pininfarina's brief was straightforward: the car would be a two-door coupé built by Mulliner, Park Ward, which shared the same platform and

Pininfarina built this two-door Bentley T which was displayed at the 1968 London Motor Show. The fastback styling is rather elegant but the headlamps seem somewhat cumbersome. It has been suggested that this car inspired the development of the Camargue. James (now Lord) Hanson bought the car but its present whereabouts is uncertain. (Courtesy Rolls-Royce Motor Cars Ltd.)

base mechanical units as the Silver Shadow. Its appointment would be even more luxurious and certainly superior to that of the two-door models currently available. Subsequently, eight sketches were prepared and Sergio Pininfarina presented his ideas to John Hollings, Geoffrey Fawn and Fritz Feller, together with members of Mulliner, Park Ward management, at a meeting in London at the coachbuilder's premises. From that meeting, a design was chosen and accepted, and the drawings and clay model were eventually prepared and delivered to Crewe.

Pininfarina's proposed styling followed that which was currently fashionable and any suggestion that Pininfarina copied the design he had employed for Fiat's 130 coupé, the two-door alternative to the Turin car

maker's top of the range model, has been denied. The suggestion of a link between the styling of the two cars appears to have originated in one of Britain's well-known motoring periodicals, where it is alleged a journalist caught a glimpse of the Delta project undergoing trials and noticed styling similarities, and also from *Road & Track* in wich Paul Frere commented that it looked: "a lot like Pininfarina's 130 Coupé". Interestingly, early sketches of the Rolls-Royce Delta, while illustrating Pininfarina's contemporary styling features, are quite unlike the definitive model.

Events at Rolls-Royce during 1971 delayed the Delta programme quite severely and it was not until the middle of 1972 that the first experimental car was made available for testing. Alto-

The Camargue was considered by Rolls-Royce to the flagship and was certainly, at the time, the most expensive car on the British market. (Courtesy Rolls-Royce Enthusiasts' Club)

This is possibly the first full-size mock-up of the Camargue. Frontal styling was particularly difficult for Pininfarina due to the mass and design of the huge radiator. The scuttle was considered too high by some. Even so, this car was especially appreciated by Rolls-Royce directors. (Courtesy Rolls-Royce Motor Cars Ltd.)

gether, six experimental cars were built, four initially and two some years later using the full SZ underframe, which was not adopted for Camargue pro-duction. The last two experimental cars also helped in preparation of the Silver Shadow's replacement, the Silver Spirit. When initially seen at Crewe, the first prototype car was something of a shock, the bulk of the vehicle quite dwarfing the Silver Shadow standing alongside. An engineer, likening it to the Rock of Gibraltar, received a curt response from a test driver who, having just driven the car for the very first time, judged it to be about as fast! In fairness, as pointed out by a Rolls-Royce stylist, the change in fashion from the subtle and delicate curves of the fifties - which could disguise the bulk of a vehicle very successfully - to the crisp sharp corners of the seventies, made even a small car look big purely by emphasis-

Several sources have suggested Pininfarina took the basic styling idea for the Camargue from the Fiat 130 Coupé. It has to be appreciated, however, that Pininfarina's then current styling trends are mirrored in this design which stems from the late sixties and early seventies. This appears to be a publicity photograph taken in 1986 for a limited edition US specification car. (Courtesy Rolls-Royce Motor Cars Ltd.)

The codename for the Camargue was Delta and the car was only ever intended as a Rolls-Royce-badged car. The styling has always provoked controversy, not least from Rolls-Royce and Bentley enthusiasts and those involved in the development of the car. The date of this photograph is unknown but, looking at the bumper assembly, especially on the car's offside, the fit might suggest this to be an experimental vehicle. (Courtesy Rolls-Royce Enthusiasts' Club)

ing its extremities. As for the interior, the mock-up was generally pleasing and was well-liked.

Testing of the Delta experimental cars was conducted using a Bentley-style radiator in common with other projects. For a time, however, the first test car was used without a radiator shell while the dummy grille was being constructed. Ironically, bearing in mind that a Bentley version of the Camargue was not contemplated, the proportions of the dummy shell seemed, to some, to suit the car's styling better than the traditional Rolls-Royce type.

A number of difficulties were experienced with the initial prototype cars: D1 (D for Delta project) suffered from excessive road noise, poor performance and the effects of side winds; D3, the second experimental vehicle (D2 was never used for road work and performed as a crash tester) was finished in a paprika colour and was not particularly liked by the experimental department which found it to be a poor performer at first. Enjoying something of a chequered career, D3 later served in the publicity campaign to launch the Camargue and was eventually scrapped with almost 100,000 miles (160,000km) recorded, but not before the car had been fitted with a 7.25-litre engine, which was subsequently re-

placed by the usual 6.75 unit. Prior to the car's destruction it was involved in a serious accident while on trials in Scandinavia.

D4, after production in 1973, was sent to France for intensive testing and John Gaskell remembers being involved in the car's 50,000 mile (80,000km) test schedule in which the only major problems were those concerning the split-level air-conditioning. One of the Camargue's features, of course, was Rolls-Royce fully automatic air-conditioning later fitted to the Silver Shadow II, having already been employed on the Corniche. Once the temperature levels had been selected, no further adjustment was required whatever the external temperature.

During the Camargue's incubation period, Macraith Fisher and Derek Coulson visited Sergio Pininfarina in Turin in readiness for the car's development programme. Pininfarina, incidentally, was very amused at the two engineers actually driving to Italy for the meeting - he was normally used to people flying in by helicopter to see him!

Like most things at Rolls-Royce, the Camargue's development programme was extremely thorough and, therefore, time-consuming. Much of the work consisted of complying with the severe emission control regulations in America, as well as that country's stringent safety requirements. The tooling process was also complicated and, like the Silver Shadow two-door saloons, building the cars was split between Mulliner, Park Ward's Willesden factory and Crewe. Pressed Steel, as with the Corniche, supplied the floorpan, dispatching it directly to Mulliner, Park Ward at Hythe Road where it was mated with the bodyshell, which itself was assembled from panels supplied not only from the MPW factory but also other sources. As with the other Silver Shadow derived models, aluminium was used for the bonnet, doors and boot lid. From Willesden the completed bodies were transported in pairs to Crewe for rust-proofing, paint priming and fitting of all major mechanical components before being returned to Hythe Road for completion and the usual extensive pre-delivery inspections and road testing.

Had the original plans come to fruition, the Camargue might have been ready for a 1973 launch. As it happened, events at Crewe dictated otherwise and, in any case, the fuel crisis of that year may well have proved detrimental to the whole project. Priced at a fraction under £30,000 with taxes, demand for the Camargue - which was almost twice the price of a standard Silver Shadow saloon and half as much again as the Corniche, was never going to be extremely high, hence the limited production facility. Launched in Sicily during January 1975, the two-door, five-seater Camargue was not only the most costly car in the whole of the Rolls-Royce catalogue, it was also the largest and heaviest Silver Shadow derivative. A whole army of motoring journalists being invited to test the publicity cars did allow the possibility of someone damaging all that expensive metalwork. Accordingly, to record the event John Hollings had a plaque prepared which he took with him to Sicily to present to the first journalist to sustain accidental damage to a Camargue. It had to happen, of course, and the hapless journalist now has a permanent, if embarrassing, reminder of the feat.

As expected, the Camargue was luxuriously appointed; the softest Nuela hide gave the utmost comfort and the seats themselves, electrically adjustable, even incorporated a powered front seat back release to enable rear passengers to get into and out of the car easily. The facia was controversial and did not appeal to everybody; instead of being constructed from wood,

the panels were made of aluminium to keep weight as low as possible, but covered in standard veneer in order to retain a traditional appearance. The instruments were surface mounted, instead of being set into or behind the dashboard - a Pininfarina touch in recognition of Rolls-Royce's long association with aviation. At the wheel of a Camargue there was every indication this was no ordinary Silver Shadow derivative: exclusive in all respects, the interior was surprisingly light and airy, the slim roof pillars helping in this regard. Of interest are the tell-tales on the leading corners of the front wings which were placed there after tests revealed that drivers were unable to accurately judge the width of the car!

From the time a platform was delivered to Mulliner, Park Ward, building a Camargue took approximately 6 months and a customer could expect to wait up to two years before taking delivery. Initially, a single car a week was supplied and, as orders increased, so delivery times extended. Part of the reason for this slow production was the difficulty experienced in obtaining suitable body panels: Rolls-Royce standards were such that much of the material was rejected on the grounds of poor quality. The Willesden works of Mulliner, Park Ward was already under severe pressure to produce Corniche models as well as the other coachbuilt Silver Shadows. To ease the situation, in 1978 assembly of Camargue coachwork was transferred from Willesden and awarded to Motor Panels Ltd. of Coventry.

Pressed Steel delivered the floorpans direct to Motor Panels Ltd. at Coventry, as did suppliers of various other body components. Once Motor Panels had built the completed shells they were passed to Crewe for rust inhibiting, priming, fitting out and road testing. The same exacting standards were employed at Crewe and there was nothing to tell the Crewe and Willesden cars apart except the maker's plate on the door sill: Willesden Camargues can be identified by plates stating *Coachwork by Mulliner Park Ward*, and the Crewe cars which read *Coachwork by Rolls-Royce Motors Ltd.* In both cases, the words *Designed by Pininfarina* left no doubt about the car's styling origins.

There were few major revisions to the Camargue's specification during its production span which continued over a period of eleven years; rack and pinion steering was fitted from 1977 and from 1979 the suspension was updated to that being developed for the Silver Spirit. In 1980, cars being exported to California were equipped with Bosch fuel injection and, a year later, the same modification was applied to all export models. From 1980 the Camargue was sold alongside the Silver Spirit range of cars as well as the Corniche. Although the Camargue never received the full SZ underframe,

it did get the rear end in 1979, just like the Corniche, but not the revised front end. In this way, the Camargue retained a ghostly element of the Silver Shadow well into the eighties.

Bentley enthusiasts were aggrieved by the fact that the Camargue was only ever specified as a Rolls-Royce. However, a single Bentley version of the car was built at a customer's special request. Car D1, however, was built as a test car in advance of a proposal to market a high performance Bentley version equipped with a turbo charger and, in this guise, was fitted with a Bentley radiator in order to be less conspicuous. Accordingly, it couldn't really be called a Bentley as it was built as a Rolls-Royce. The project never materialised as such and the car was ultimately scrapped but not before some incredible results had been recorded, including a top speed of 145mph (232km/h). As for the specially ordered Bentley Camargue, certain styling difficulties were experienced in mating the radiator shell to the bonnet, as the same pressing as that for the usual Rolls-Royce-badged car was used.

In total, 526 Camargues were built which assures the model's exclusivity. The last cars were built in 1986 with the final batch of 12 cars exported to the USA during the early period of

1987. Not everybody approved of the car's styling and history shows it to be one of the most controversial Rolls-Royces but, most certainly, one of the world's great cars.

A few words about the latter-day activities of Mulliner, Park Ward are in order before closing the chapter. The economic climate, as well as a demand for the true coachbuilt car, meant that by the mid-nineties it was no longer viable to retain the company's London premises and closure was announced in May 1991, whereupon building of the Corniche, Bentley Continental and Silver Spur Limousine was gradually transferred to a dedicated site at Pym's Lane. The Phantom VI was discontinued in 1992 and the workforce at Willesden cut back accordingly to just a handful of craftsmen in mid-1993. With work on the Touring Limousine (which is built on an extended Silver Spur platform) concentrated at Crewe, the famous coachworks in Hythe Road, Willesden closed its doors for the last time in August 1994.

The spirit of the Shadow lives on, however, and the very last Corniche has a place of honour within the Gallery at Pym's Lane. As for all the other coachbuilt derivatives of the Silver Shadow and Bentley T-Series cars, their futures are also assured. Enthusiasts of the marques will see to that!

ROLLS-ROYCE & BENTLEY

V

LIVING WITH A SILVER SHADOW OR BENTLEY T-SERIES

There is probably no product anywhere in the world as marketable as a Rolls-Royce. Even the Bentley, with its famous winged-B mascot, cannot match that of the famous Double-R emblem or the majesty of the Spirit of Ecstasy. Whether one appreciates the marque or not, no other car is as instantly recognisable as a Rolls-Royce, and certainly no other car has the ability to make such an impression. Rolls-Royce as a company has, therefore, a formidable reputation to maintain; one revered for something approaching 100 years which is synonymous with excellence.

To a great many motorists, the idea of ever owning a Rolls-Royce or Bentley motor car, however highly they regard the marques, is beyond all contemplation. Others, however, in acknowledgment of the cars' fine engineering and exquisite comfort, together with the company's commitment to quality, will accept nothing less. This attitude may be true of buying a new car, but what about purchasing an older model, one that has become a recognised classic? The Silver Shadow and Bentley T-Series cars could, for many motoring enthusiasts, represent the ultimate classic car.

Without any doubt, the Silver Shadow and its Bentley equivalent have become recognised classics in their own right. For potential buyers there is good news in as much that the survival rate of these cars is extremely high, prices are genuinely affordable and, very importantly, the supply of replacement parts is excellent. Before readers rush to buy the car of their

Sculpted in 1911 by Charles Sykes ARBS, the Spirit of Ecstasy adorns some of the greatest cars in the world. (Author's collection)

dreams, they should consider that there is another side to ownership: running costs can be exceedingly high, some parts are hugely expensive and, unless the owner is an experienced mechanic, servicing and repairs are often beyond owners' capabilities. Contrary to popular belief, the Silver Shadow range of cars require just as much maintenance as any other motor vehicle and therefore, due to engineering complexity,

Now classics in their own right, the Silver Shadow range of cars has survived in exceptionally large numbers. There are also many specialists who supply and maintain these fine cars. (Courtesy Bill Bateman)

will probably require specialist attention sooner or later. Once having accepted the 'for and against' argument, ownership of these cars is a practical proposition and one which can be both a wholly rewarding and enjoyable experience.

Buying advice

Due to the number produced and an excellent survival rate, there's no shortage of cars for purchase (although the T-Series Bentley is not as widely available as less were built).

There are four obvious markets from which to purchase a car: private sale; a Rolls-Royce franchised dealer or recognised marque specialist; a second-hand (Rolls-Royce prefers the term 'previously-used') car dealer, or at auction. Dealing first with the last-mentioned option, only a person experienced at purchasing vehicles at a car auction should choose this route and certainly it is recommended the buyer has some knowledge of what to look for on a particular car. It may be, of course, that a potential owner could arrange for a Rolls-Royce and Bentley specialist to undertake this task on their behalf. Similarly, when contemplating buying from a used-car dealer, unless the company has a high reputation, anyone considering a purchase in this respect is particularly advised to seek specialist advice. Buying privately can be fraught with difficulties unless the potential purchaser has some experience of these cars. When viewing a car it is essential to ensure it is without damage and that it has a full (preferably R-R) service history. If the history

of the car is not available it might well be prudent to have an independent appraisal as to its overall condition. It should be understood that if any major mechanical repairs are necessary, these could well be expensive to put right, especially considering that the cost of an engine overhaul will run into four figures. The same goes for the

bodywork; however insignificant any damage might appear, it will undoubtedly be costly to repair. Always check for coachwork repairs: unless carried out expertly these can usually be detected by either poor quality repainting and/or badly fitting panels with uneven gaps between them. A car which has been involved in an accident and

If there is evidence of coachwork repair, always ensure it has been carried out properly. When contemplating buying a Silver Shadow it is wise to obtain a specialist's report on the car beforehand. (Courtesy Michael Hibberd)

not properly repaired could be dangerous.

A safer approach - especially for the enthusiast considering their first Rolls-Royce or Bentley - could be to buy a car through a recognised dealer or at least a marque specialist who will be happy to offer the customer a comprehensive service. A franchised Rolls-Royce agency has not only the entire support of the manufacturer, but also all the technical information pertaining to a particular car. Marque specialists will also be able to offer a service similar to that of a Rolls-Royce dealer but, in either case, it would be wise for anyone seeking a car to discuss their requirements beforehand and to ascertain the future support a company can provide.

It has to be appreciated that, at the time of writing, the last Silver Shadows produced are at least 16 years old and early examples have been around for up to 30 years. Some Rolls-Royce dealers are therefore reluctant, unless the car is particularly special, to offer such vehicles for sale and prefer, instead, to sell the car on through the motor trade to an independent specialist. That, of course, does not preclude them offering specialist servicing and advice.

It goes without saying that buying through a specialist will almost certainly be more costly than purchasing at auction or by private sale. As a measure of safety, however, it is hoped that the intended purchase would be delivered in an excellent condition and, in the long term, represent not much more expense. In broad terms, a Silver Shadow or T-Series Bentley in good condition will probably be no more expensive to buy than an average popularly-priced new car.

When looking at different cars, don't be too put off by the vehicle having changed hands a number of times. Initially, quite a number of cars were sold to companies and, usually, sold on after a couple of years. Some owners bought new cars at regular intervals - every two to three years, usually - because during the sixties and seventies it was possible to sell a car for the same amount, and often more, than originally paid, so making the purchase of a new car very advantageous. As long as the vehicle has been well cared for and service schedules adhered to, there should be little problem.

Potential as well as existing Silver Shadow and T-Series owners will derive much benefit from joining the Rolls-Royce Enthusiasts' Club or the Bentley Drivers' Club, and it is through these organisations that marque specialists can be sourced. The R-REC does have at its disposal the complete build histories of every car and often

particular vehicles are known within club circles.

Prices for Silver Shadow and T-Series Bentley saloons can vary greatly depending upon age and condition of the car. Coachbuilt models often command even higher prices, due to appeal and rarity factor. So, what should the potential purchaser pay for a car? According to Rob Jones of Benver Services, an independent Rolls-Royce specialist in Crewe, the minimum anyone could expect to pay for a Silver Shadow is around £3000. For that sum, the car would be in need of complete restoration which, as well as being time-consuming, would also be very costly, certainly far more expensive than buying a car in first class order. For a car costing something like £5000, expect to pay a further £10,000 to achieve a reasonable condition. John Bowling of Bowling-Ryan, a Rolls-Royce specialist in Bolton, Lancashire, would not contemplate buying a car under £7500 as anything less will require extensive restoration. For a superior first-series Silver Shadow/Bentley T, he advises it should be possible to pay somewhere around £10,000, although Series II cars and those vehicles in absolute pristine or concours condition may command considerably higher prices.

A lot of people, when considering buying a Silver Shadow today, do not propose to use it as daily transport but keep it, instead, as a classic car to be driven on high days and holidays. That does not mean the car's unsuitable for daily use although high mileages will obviously be expensive. Despite this, it does make for delightful daily trans-

A two-door Saloon/Convertible or a Corniche model will, because of desirability and low build quantities, attract higher prices than the standard saloon cars A car in firstclass condition can command prices of up to £30,000. (Author's collection)

port - as a great many owners know.

Corniches and their predecessors, the two-door saloons and convertibles, along with the long wheelbase cars, make an attractive proposition. Due to rarity factor alone, prices of the Mulliner, Park Ward cars are higher than for the Standard Saloons and cars in first class condition can command up to £30,000. A good example of a Camargue, even accounting for the car's controversial styling, can attract prices as high as £55,000 but it is possible, as with the Cornichc models, to obtain cars in a lesser condition for around a third of that price. Remember, though, any restoration of a coachbuilt variant is likely to be very, very expensive. As for the long wheelbase models (and that includes the Silver Wraith II) these cars not only provide extra comfort for rear seat passengers, but attract generally similar values to those of the Standard Saloons. Be warned, though, the extra length might mean it won't fit into your garage ...

Prices of the much rarer Bentley T and T2 are, surprisingly, no higher than for the Rolls-Royce-badged cars and, according, to current published prices of classic cars, early models tend to be even cheaper. The Bentley does have a particular following, however, and many dedicated enthusiasts are keen to promote the marque's sporting image and traditions. There are those motorists, of course, who admire the engineering finesse of the Silver Shadow and T-Series cars, yet prefer the softer frontal styling treatment afforded to the Bentley.

Production of the Corniche and its Bentley sister of latter days, the Continental, continued until 1995. Values for these cars, which were built from the early eighties onwards, will be somewhat greater and it can be safely assumed that the Corniche II, III and IV will carry very high prices.

For anyone considering the purchase of a prestigious car as a classic vehicle, it's interesting to compare prices with alternative marques. For an Aston Martin DBS V8 of 1969-73 vintage, one can expect to pay up to £18,000 for a good example, and even more for a V8 from 1969-73. Daimlers and Jaguars, both of which enjoy a loyal following, carry prices of only a little less than those of the Silver Shadow, while a Mercedes-Benz 450 can command much the same.

When considering the purchase of a T-Series or Silver Shadow, it is important to understand in what sort of condition a car might be for the amount of money asked. A vehicle requiring complete restoration is unlikely to be the choice of a novice to the marque as specialist equipment will be needed to perform the work. Apart from the mechanical aspect, the bodywork and, more than likely, interior, would also require attention. A more expensive car in better all-round condition may still incur some expense. Should the car require repainting, expect to pay at least £5000-6000 for a reputable job to Rolls-Royce standards; interior work to the leather or wood trim may also be expensive, especially if original materials have to be matched. Remember, also, that items such as tyres and

external trim easily get overlooked in the enthusiasm to buy a car; tyre wear on earlier cars was always heavy and more than 15,000 miles (24,000km) from those vehicles with non-compliant suspension is unusual. Tyres are expensive: a set of four can cost in excess of £700 and that does not include fitting, wheel-balancing or local tax.

Should the bumpers on Series I cars have sustained even minor damage, the cost to replace them, front and back, will be around £2200. Even a corner-piece section will amount to almost £250. The bumper assemblies on Series II cars are hugely more expensive and cost as much as £2040 for a front replacement and £1950 for the rear section.

Replacement exhausts cost around £600 for the genuine article and it is false economy to fit anything other. Non-Rolls-Royce systems may appear a little cheaper but usually will not fit or wear as well, which means earlier replacement and even more expense. Take a look to see whether the car's tool kit is in place and, if so, its condition. The fact that it is missing or in a damaged state could tell a lot about the vehicle's history ...

Ideally, when contemplating buying a car, remember that, as far as Rolls-Royces and Bentleys are concerned, people are normally prepared to pay for quality; cars in lesser condition usually sell at a price which reflects the amount of money needed to put it in good order. A car will retain its quality long after the initial cost has been accepted and forgotten. The ad-

Silver Shadow Saloons 1965-71	£4500-9000
Silver Shadow Saloons 1972-76	£7000-12,000
Silver Shadow Long Wheelbase	£9000-13,000
Silver Shadow MPW FHC	£12,000-18,000
Silver Shadow MPW DHC	£14,000-22,000
Silver Shadow II Saloons 1977-79	£8000-16,000
Silver Shadow II Saloons 1979-81	£12,000-24,000
Corniche FHC 1971-77	£14,000-27,000
Corniche DHC 1971-77	£15,000-40,000
Corniche DHC 1976-88	£25,000-70,000
Camargue FHC	£25,000-50,000
Bentley T Saloons 1966-72	£9000-13,000
Bentley T Saloons 1972-76	£9000-14,000
Bentley MPW	£12,000-22,000
Bentley T2 Saloons 1977-79	£12,000-16,000
Bentley T2 Saloons 1979-81	£15,000-25,000

vice, therefore, is to pay as much as can be afforded for a car in good condition and with a suitable history. John Bowling's advice is to opt for a Silver Shadow II or Bentley T2 because of the cars' refinements over earlier models. The handling on these cars is much improved on that of the first series and benefits from having rack and pinion steering. The fully automatic air-conditioning is far better than the system employed on the first series cars and provides greater comfort. Also, once set, it never needs to be altered.

There are, of course, a great number of first series cars in day-to-day use, many of which are superbly maintained and in splendid condition. Some enthusiasts actually prefer the earlier cars to later models and these cars will, naturally, provide the potential owner with a first class investment. John Bowling always advocates the use of radial-ply tyres on early cars as they provide better road-holding, even at the expense of greater road noise.

John reckons potential purchasers often place too much importance on low mileage cars. While this may seem a plus at the time of buying, it could mean that the vehicle has been used mainly for short journeys, a possible cause of excessive engine wear. There's also the possibility that components could fail once the car is subjected to greater use.

Most official Rolls-Royce agents, and independent specialists, will offer to survey a car and supply a written report on condition before purchase. Obviously, some agents and special-ists make a charge for this service but John Bowling's policy is to provide a free report in the interests of customer satisfaction. Unless the purchaser is confident he knows exactly what to look for on a car, it would not be sensible to buy a car without first having a survey, which could save a lot of disappointment and later expense.

For insurance purposes, the Rolls-Royce Enthusiasts' Club publishes the above list of current (1996/7) values, depending on condition.

When looking at cars for sale, it is important to be able to determine at a glance its approximate age. This is not so easy with a Silver Shadow or Bentley equivalent because, throughout the 15-year production span, very few styling changes were implemented. To the unwary, a late Silver Shadow II or T2 can look remarkably similar to an early first series car. If the car has plastic-faced bumpers and an air dam, it is a Series II model; cars with chrome bumpers and overriders are from the first series and those without grilles beneath the headlamps are identified as late versions. Finer identification is

possible from the type of facia the car has: early first series cars have a large expanse of woodwork without a central console, while those cars complying with US Federal safety standards have greater amounts of padding, together with a V-shaped console. Telling a car's age is a lot easier with home market cars than those exported to the USA. American cars were never fitted with an air dam and even first series cars had energy absorbing bumpers from 1974. To be sure about a car's age, of course, the chassis number provides positive identification.

Finally, unless buying privately or at auction, always purchase from a reputable source, even if a higher price has to be paid. Avoid a car that has been neglected or has a suspicious history - it is always easy to be wise - and sorry - after the event.

A Shadow in the garage

To the unwary, an early Silver Shadow example could be mistaken for a later car. Styling features such as bumpers and the design of the facia are a good way of establishing a car's age, but the surest way is by the chassis number.
(Author's collection)

Before thinking of driving the chosen Rolls-Royce or Bentley home, the new owner will obviously have checked that the car will actually fit inside his or her garage ... There must be at least one enthusiast who's had the problem of enlarging the garage which probably cost more than the car!

The satisfaction derived from owning a quality car with one of the most famous emblems of any marque is, however, just the tip of the iceberg. To drive it and use it to full potential, even a Rolls-Royce or Bentley needs regular servicing and preventative maintenance. Having purchased a good example Rolls-Royce or Bentley for the same money as an economically-priced new family saloon, the running costs are, quite definitely, still within the Rolls-Royce league.

Expect no more than 12-18mpg (24-16lts/100km) in fuel consumption depending upon engine condition, as well as the overall state of the car. Servicing costs for a Silver Shadow and T-Type can also be higher than for some of the more popular makes of car but, compared to a lot of 20-30 year old cars, there is no problem concerning availability of parts. Current servicing schedule prices show relatively little difference between official Rolls-Royce dealers and independent specialists: for example, Lancaster Europa Ltd., a Rolls-Royce and Bentley agent in Shef-field, quotes £175, £350 and £650 for 6 months/6000 miles (9600km), 12 months/12,000 miles (19,200km) and 24 months/24,000 miles (38,400km) services respectively, including parts, labour, tax and a free MoT test. Bowling-Ryan, however, charges £188, £323 and £588 for the same work. Usually it takes 1, 2 and between 2-3 days to carry out the respective service schedules. In addition, 48,000 and 96,000 mile (76,800 and 153,600km) services are major affairs, usually entailing the vehicle being off the road for a week. Obviously, these services are expensive and, at the time of writing, minimum cost is somewhere in the region of £1000. These prices may appear fairly acceptable for a Rolls-Royce but, remember, they represent only routine servicing and do not account for any other repairs or maintenance that might be required. For example, should the hydraulic system need a complete overhaul, expect a specialist to tot up a labour time of around 50 hours to change all the seals.

When asked what made working with Rolls-Royce and Bentley cars different to any other, John Bowling was adamant it was the quality of the original materials, together with engineering excellence. Even replacement parts, he added, are developed to the same high standard. Like official dealers, John Bowling will not fit anything other than Rolls-Royce parts and confirmed that, in many instances, these were no more expensive than non-Rolls-Royce items. John also emphasises the need to keep strictly to manufacturer's servicing schedules. Taking short cuts is no cheaper in the long run and can result in unreliability.

Rolls-Royce has an excellent replacement part service. Reg Vardy plc, official dealer for Tyne and Wear, can normally order any item up to 4.30pm for delivery by the time the dealership opens for business the following morning. In cases of urgency, customers are safe in the knowledge they can leave their car overnight and it will be ready for collection early the next day. It is not only within the United Kingdom that the company supplies replacement parts: consignments leave the warehouse by road three or four times a day en route to home-market distributors, the European regional warehouse in Switzerland, to Lyndhurst, in New Jersey, which serves the American continent and, of course, almost any other destination worldwide. In cases of urgency, parts are air-freighted on the first available flight.

To neglect a Rolls-Royce or Bentley and miss important services is, ultimately, devaluing the car and, of course, could result in MoT failure. Even worse, neglect will pose a safety risk. All cars, including the products from Crewe, wear out eventually or need adjustment, and proper service schedules must be adhered to if the owner intends keeping the car in the condition it deserves. Not even a Rolls-Royce is free from the possibility of

Rolls-Royces do sometimes require extensive attention ... (Courtesy John Bowling)

Right: There are certain areas on a Silver Shadow which should be checked before and after buying. The wheelarches are prone to corrosion due to ingress of moisture. (Courtesy Michael Hibberd)

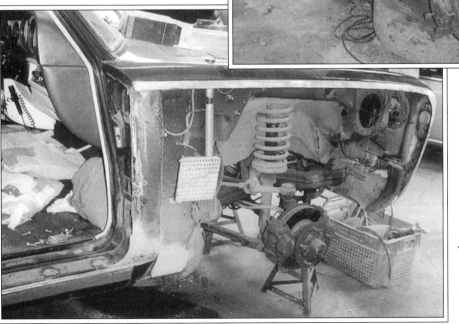

Left: In this instance neglect has resulted in severe corrosion of the area around the front wheelarches, front wing lower sections, inner wheelarch filler panels and outer sills. The wing assembly has been removed to facilitate replacement and repair. (Courtesy Michael Hibberd)

133

Left: The boot floor is another area which should be checked periodically. Always use genuine Rolls-Royce components when making repairs: the cost of a stainless steel exhaust is around £600 but it will fit perfectly, whereas a non-original system will not and will require replacement sooner. (Courtesy Michael Hibberd)

Right: All drain holes should be kept clear and the sills regularly inspected. Doors are made from an aluminium alloy and, being softer than steel, are more prone to accidental damage. (Courtesy Michael Hibberd)

Left: Leaded joints on older cars are susceptible to deterioration which should be rectified as soon as detected. Tyres and brake pads tend to wear quickly; rear brake discs are prone to corrosion and handbrake failure is not unknown. (Courtesy Michael Hibberd)

corrosion, however, and the Pressed Steel bodies, which were the product of mass-production, can eventually show signs of rust, although the effects of rot do take longer to appear than on most othercars. Preventative maintenance is, therefore, a very serious part of owning a Silver Shadow or Bentley T-Series.

As with all cars, the T-Type and Silver Shadow do have weaknesses. Bodily, the first signs of decay are usually to be found around the wheelarches, the back edges of the front and rear wings in particular. This is a common problem and is due to the ingress and retention of water. It's essential to ensure that all drain holes are kept clear. Michael Hibberd, a Rolls-Royce specialist in Slough, Berkshire, advises these should be checked regularly as he has known of cases where as much as a couple of gallons (9.5 litres) of water have been drained from the sills. In extreme cases, the floor of the car has rotted away entirely. It is prudent, therefore, to examine the carpets and underfelt as they act like a sponge, quietly rotting the metal beneath. Corrosion in this region is best repaired at once rather than leaving it to get worse. Regular inspections around the car are always advisable and periodic checks of boot floor condition and chrome trim mouldings will prevent minor corrosion from going unnoticed.

The doors on Silver Shadows and T-Types are made of aluminium alloy and the effects of many winters' salted roads can cause corrosion; also check that salt has not affected the brightwork, door handles and locks. Being of light alloy, and therefore softer than steel, doors are susceptible to even slight accidental damage. The same, naturally, goes for the bonnet and boot. Other areas to watch, especially on older cars, are the leaded joints which can deteriorate; initial evidence of corrosion in this respect will be some cracking of the paintwork.

Coachwork quality of the Mulliner, Park Ward cars is generally considered not much better than that of the Standard Saloons - an indication of just how well the Crewe product was constructed. This really is not surprising as the Standard Steel cars of the pre-Silver Shadow era received the same attention as their coachbuilt versions.

Mechanically, there are several points to check: rear brake discs can corrode, usually because of inactivity, and handbrake failure is not unknown because the mechanism operates on the edge of the brake disc. Brake pads tend to wear fairly quickly and, depending upon the use expected of them, may have to be replaced at intervals as short as 10,000 miles (16,000kms). Ball joints, especially on cars with compliant suspension, are susceptible to wear, if only because of the enormous amount of weight placed on them.

The hydraulic system, although well-proven, can give rise to problems. Any repairs here can be costly so it's best to investigate any troubles as soon as they occur. With the engine started, the car should rise to its normal height very quickly and changes in weight distribution within the vehicle should be accounted for almost instantly. Any leaks from the hydraulic system, and that includes the power steering pump, should, of course, be investigated immediately.

The engine, whether the original 6.2-litre or later 6.75-litre unit, is very reliable and should, if treated with respect, last for at least 100,000 miles (160,000km) without causing concern. Remember, though, from the driving seat there will be little evidence of any malfunction due to the unit's inherent smoothness and effectiveness of the vehicle's sound-proofing material. Telltale signs of impending trouble may be a smoky exhaust and increase in oil consumption, as well as unacceptable sounds from the valve gear. If the car has been idle for any length of time, do, however, expect some initial noise from the engine until the oil has had time to pressurise the hydraulic tappet mechanism. If a knocking sound persists, suspect corrosion of the piston liners. Don't be surprised to see a drop of oil on the driveway or garage floor: this is a common feature and one enthusiast reckons that most Silver Shadow owners have a gravel parking area because of this! One reason for this leak is that there is no rear oil seal on the crankshaft, and the problem is made worse if the car is parked facing uphill. Another reason is that the sump is often over-filled with oil. With age, gaskets may deteriorate and allow some seepage of oil to drip down the outside of the block. Obviously, this should be rectified at once.

The engine on Silver Shadows and T-Series cars is not as quiet as is often portrayed and certainly the old claim

As long as it is well-maintained, the engine of a Silver Shadow should provide at least 100,000 miles (160,000kms) service without major attention. The same applies to the gearbox and final drive. (Author's collection)

that the only sound heard at 60mph (96km) is the ticking of the clock is not altogether correct. Under normal circumstances the engine is audible but, naturally, not intrusive.

The transmission is usually reliable; the GM400 gearbox is a superb unit which rarely gives trouble. Most cars around will be fitted with the 3-speed box and it will only be the very early home market cars that have the old 4-speed unit. Normally, the GM400 gearbox can provide up to at least 100,000 miles (160,000km) without any need for overhaul, the most common fault being a leak from the front pump seal. 4-speed boxes may need attention after 80,000 miles (128,000km) and the standard telltale sign of wear is usually uneven and clumsy gear changing.

The rear axle is also very reliable; early cars did suffer from some vibration but the problem was soon corrected. The unit should cover at least 100,000 miles (160,000kms) without major attention. Wheel bearings are also reliable and are likely to give very long service.

Front suspension bushes should be regularly checked, especially around compliant mounts and anti-roll bar supports. Thuds from the rear end of the car - especially when pulling away - can indicate worn torque arm rubbers on the rear-drive universal joints.

On vehicles as mechanically complex as the Silver Shadow and T, it's understandable that the majority of owners entrust their cars to a skilled specialist. There are enthusiasts, however, who do carry out their own servicing and maintenance and it has to be said that the Rolls-Royce Enthusiasts' Club organises for their members workshop events to explain the mechanical aspects of the cars. It's not possible to cover all aspects of Silver Shadow maintenance here, but the following are some of the more important servicing requirements of which the owner should be aware.

The engine and hydraulic systems are the two areas of a Silver Shadow and Bentley T that warrant meticulous attention. Engine oil and filter changes should be carried out strictly in accordance with manufacturer instructions, and the correct fluid used in the car's hydraulic system. There are also associated requirements such as maintaining the cooling system, checking and - where necessary changing - drive belts, attention to the braking system and changing brake pads when required, examining tyre condition and exhaust system: in other words, spending a little time, on a regular basis, on preventative maintenance.

It has already been mentioned that maintaining and repairing a Silver Shadow or T-type is probably beyond a lot of owner's capabilities. It cannot be stressed enough the folly of attempting any work on these cars without full knowledge of procedure and use of special tools where required. Most certainly a lack of experience can be extremely dangerous and it would be very unwise to attempt a do-it-yourself restoration if this is the case.

Behind the wheel

The Silver Shadow and T-Series Bentley may be good to look at but it isn't until one is behind the wheel that the cars can be fully appreciated. The sumptuous interior, with its top grade leather upholstery, deep pile carpeting and exquisite burr walnut veneer, is the height of good taste and complements the car's sophisticated engineering. Once settled into the comfort of the seats - electrically adjustable, of course - which provide a commanding driving position, a feeling of relaxed well-being is inevitable.

As the ignition is turned on and the engine started, the sound of the V8 power deep under the bonnet is immensely satisfying. As the controls come alive the self-levelling suspension immediately raises the car to its correct running height whilst power steering makes directing the car effortless. A flick of the gear selector is all that's necessary to start the car in motion; release the handbrake whilst applying the gentlest pressure on the accelerator and you're away with a reassuring surge.

Even in Bentley guise, the Silver Shadow cars are far from performance machines and are most unsuitable for motor sport. Only the brave, therefore, would expose such a heavyweight car with all its finery to the rigours of the rally course or track. A Silver Shadow was, however, entered in the *Daily Mirror* World Cup Rally of May 1970 for which it was extensively modified by its owner, Ray Richards. In addition to uprated suspension and a safety bar over the front of the car, extra driving lamps were fitted and a roof rack carried extra tyres and wheels. Rather

Not all owners would want to take a Silver Shadow rallying! Ray Richards entered his car in the 1970 Daily Mirror World Cup Rally. This is a specially-prepared publicity photograph, taken not on the rally but in a quarry!
(Courtesy National Motor Museum)

bizarrely, the exhaust pipes were routed away from the vehicle's underside, through outlets in the bonnet and over the top of the car! Richards had hoped for support from Rolls-Royce for his venture but when this was not forthcoming he purchased an American specification car with left-hand drive. The car did not perform as well as anticipated and, in order to increase power, several modifications were carried out to the running gear and exhaust. Still the car lacked power and it was only after a cry for from the owner that the manufacturer became involved in the project.

With Rolls-Royce to the rescue, the engine and exhaust were adjusted to specification settings, and a problem concerning overheating of the hub assemblies was investigated. It appears the bearings were packed with a grease with too high a melting point which caused the hubs to run dry, so resulting in damage. The car was eventually prepared in time for the rally but it seems the going proved rather tough and Ray Richards eventually arrived at the Lisbon stage in 63rd place after suffering transmission problems. Grease had leaked from the driveshafts due to the rough conditions, the bearings seized and the shafts snapped.

A local Rolls-Royce dealer responded to the situation and cannibalised a car he had in stock which allowed Ray Richards to continue to the next rally stage. It was in the South American section when, heavily grounding, the driveshafts were torn away from the final drive unit and the car had to retire from the rally. Nevertheless, the Silver Shadow was repaired and used as a support car for the remainder of the rally course.

T-Series Bentleys have been used for hill-trial events at Prescott in a demonstration capacity staged by Rolls-Royce but, more latterly, a Silver Shadow has taken part in the Land's End to John O' Groats Reliability Trials (LE JOG). However, competitive motor sport really is not what the car's designers had in mind, and the Silver

137

Shadow is far better equipped to be a very capable tourer, affording the highest degree of refinement and comfort.

The number of modifications made to the model throughout its production life mean that the late cars are rather different in feel and handling. Rolls-Royce accepted there were some shortcomings with handling, hence the modifications that were introduced. Series II cars have a more stable ride than earlier models and the rack and pinion steering is also more precise than the original recirculating ball device. In fairness, however, the Silver Shadow was never intended to have out-and-out performance but rather superior ride quality, although some owners of the first batches of vehicles experienced rather unpleasant sensations, akin to sea-sickness, due to the car's soft suspension.

In recognition of its shortcomings, Rolls-Royce eventually offered a handling package to improve roadholding and stability. Should a car's suspension system be showing signs of wear, the fitting of a handling kit is well worth consideration as it much improves the behaviour of the car. Taking into account the cost of replacing a set of dampers and springs, fitting a handling package costs something like £1200, plus about £400 labour charge (these figures do not include tax) makes economical sense.

An alternative to Rolls-Royce's own package is that devised by Harvey Bailey Engineering (HBE). Designed to cut out the pitch and roll usually associated with the Silver Shadow range of cars, the Harvey Bailey kit makes the car perform as a vehicle of its size, weight and power output should. Many owners have already fitted the Harvey Bailey modification to their cars and the general opinion is that handling has been transformed. The Harvey Bailey kit consists of specially designed front and rear anti-roll bars, as well as redesigned and uprated springs. Harvey Bailey Engineering also recommends the fitment of Bilstein shock absorbers, which it says provide greater damping ability along with enhanced ride quality. The cost of the Harvey Bailey kit is fractionally cheaper than the Rolls-Royce product, current prices being £1150 unfitted and £1450 with fitting plus taxes where appropriate. The HBE kit was developed by Rhoddy Harvey-Bailey along with his business partner Graham Martin. Racing and test driver, as well as Rolls-Royce devotee, Rhoddy designs high-performance chassis and suspension layouts for a number of manufacturers and vehicle specialists. An enthusiast with experience of Silver Shadows and Bentleys 'before and after' the fitting of the Harvey Bailey kit is Andrew Morris. His beautifully kept T2 now has exemplary handling and road-holding which is as good as, if not better than, any other large high performance saloon. He well remembers, however, driving a Silver Shadow without modification, and the car's lumbering motion while cornering and negotiating bends.

Some owners do not feel the need to fit either of the handling packages, whilst others claim they make 'The Best Car In The World' even better. Bill Allen, the stylist who created the Silver Shadow two-door saloons and Corniche, did make the point, when asked about the handling, that the cars were prepared very much with the American customer in mind; also that, at the time of development, it was difficult to achieve superlative ride quality with high performance.

After driving a Bentley T or a Silver Shadow for a little time, one gets thoroughly spoiled by the car's effortless performance, unashamed luxury and sophistication. The later cars *do* have better handling, *are* better equipped and *will* afford greater driver and passenger comfort, especially with the refined air-conditioning. That's not to say that owning a Series I car is any less rewarding than owning a Series II, as each car has a special personality of its own.

For more of a sporting experience the Corniche and especially the Convertible - may be the ultimate Rolls-Royce or Bentley. These cars, with their hand-built coachwork, exclusive interior incorporating a modified facia layout and wood-rimmed steering wheel, enjoy a charisma all their own and owners rejoice at being able to cruise with the hood lowered. Open-air motoring is fine in California or on the Riviera, but in some countries the climate makes it a luxury which can rarely be enjoyed.

The unique styling and hand-crafting of the Mulliner, Park Ward cars - and that includes the long wheelbase models - does have particular appeal and driver satisfaction is great. The same goes for the Camargue which, with distinctly different styling, has a

Some enthusiasts want a more sporting approach to motoring than a saloon can offer and a Corniche Convertible could be the answer. (Courtesy Rolls-Royce Enthusiasts' Club)

unique attraction.

To increase enjoyment of the Silver Shadow and T-Series cars, many owners will want to join one, or both, of the recognised clubs, addresses for which can be found in the appendices. Both the Rolls-Royce Enthusiasts' Club (which equally welcomes Bentley owners) and the Bentley Drivers' Club offer unique membership benefits and go a long way towards ensuring enjoyable ownership of these fine cars. The Rolls-Royce Enthusiasts' Club was formed in 1957 and now has its headquarters at The Hunt House, Paulerspury, Northamptonshire. Along with an exhibition area and a meeting room, all of Rolls-Royce's extensive vehicle archives are housed within a specially built library. Not only is it possible for a member of the club to obtain the detailed build history of a particular car, but research can also be undertaken. The R-REC headquarters is a mecca for Rolls-Royce enthusiasts worldwide and during, any year the staff welcome many new owners. The club, which has a number of regional sections within the United Kingdom, as well as sections established throughout Europe and Canada, has currently around 8000 members. A superlative magazine is produced bi-monthly and, in addition, members receive a monthly advertiser detailing many specialist services and cars available.

The Bentley Drivers' Club, which has its headquarters at Long Crendon, Buckinghamshire, also welcomes owners of T-Series Saloons and Corniches. The BDC currently has 3000 members worldwide; there are 8 regions within the United Kingdom and approximately 30 overseas. Members receive the Bentley Drivers' Club Review, which is published quarterly, as well as a bi-monthly advertiser and monthly newsletter.

Whichever model Silver Shadow or T-type an owner has (and a number of enthusiasts have more than one) one thing is certain: there's no other car quite like it ...

APPENDIX I
PRODUCTION FIGURES

ROLLS-ROYCE	
Silver Shadow	16,717
Silver Shadow Long Wheelbase	2780
Silver Shadow II	8425
Silver Wraith II	2135
Silver Shadow MPW Saloon	568
Corniche Saloon	1108
Silver Shadow James Young Saloon	35
Silver Shadow MPW Convertible	505
Corniche Convertible	3239
Corniche II	1234
Corniche III	452
Corniche IV	219
Corniche S	25
Camargue	529
TOTAL ROLLS-ROYCE CARS BUILT	**37,971**

BENTLEY	
T	1712
T Long Wheelbase	9
T2	558
T2 Long Wheelbase	10
T MPW Saloon	98
Corniche Saloon	63
T MPW Convertible	41
Corniche Convertible	77
Continental	433
Continental Turbo	8
T James Young Saloon	15
T Farina Coupé	1
Camargue	1
TOTAL BENTLEY CARS BUILT	**2585**
TOTAL (ROLLS-ROYCE & BENTLEY)	*40,556*

MULLINER, PARK WARD COACHBUILT CARS

MODE	DATES PRODUCED	RHD	LHD	CHASSIS START NOS.
Rolls-Royce				
Two-door saloon	1966-71	369	199	CRH 1148
Corniche saloon	1971-76	365	274	CRH 9770
Corniche saloon	1976-77	44	97	CRH 22648
Corniche saloon	1977-79	58	95	CRH 30011
Corniche saloon	1979-80	68	107	CRX 50004
Convertible	1967-69	109	137	CRH 1698
Convertible	1969-71	124	135	DRX 6646
Corniche	1971-76	388	587	DRH 9919
Corniche	1976-77	48	206	DRH 22583
Corniche	1977-79	49	223	DRH 30003
Corniche	1979-81	116	321	DRH 50003
Corniche	1981-88	134	1167	BCX 01903
Corniche II	1986-89	60	1174	GCX 13162
Corniche III	1990-91	71	380	LCH 30001
Corniche IV	1992-93	24	90	NCX 40001
Corniche IV	1994-95	11	94	RCX 50001
Corniche S	1995	-	25	SCX 50086
Bentley				
T Two-door saloon	1966-71	79	19	CBX 1149
Corniche	1971-76	27	16	CBH 10420
Corniche	1976-77	2	1	CBH 24209
Corniche	1977-79	6	1	CBH 31226
Corniche	1979-80	4	6	CBK 50037
T Convertible	1967-69	24	7	CBH 3049
T Convertible	1969-71	8	2	DBH 7124
Corniche	1971-76	31	7	DBH 10122
Corniche	1976 -77	4	1	DBH 24505
Corniche	1977-79	1	6	DBG 31219
Corniche	1979-81	2	13	DBK 50042
Corniche	1981-85	3	9	BCX 02499
Continental	1985-89	43	134	GCX 13412
Continental	1990-91	67	113	LCX 30002
Continental	1992-93	8	33	NCH 40002
Continental	1994-95	5	30	RCX 50003
Continental Turbo	1992	3	-	NCH 400491
Continental Turbo	1995	-	5	SCX 50140

At-a-glance chronology

1965 .... Silver Shadow and T-Series launched at Paris Motor Show in autumn; a couple of weeks later the models had their British debut at the London Motor Show.

1966 .... Two-door saloon introduced by James Young. Shortly afterwards the Mulliner, Park Ward Two-door saloon was introduced.

1967 .... Mulliner, Park Ward Convertible introduced. Pilot batch of 10 long wheelbase saloons constructed, one of which was specially built for HRH Princess Margaret.

1968 .... GM400 automatic gearbox standard on all cars.

1969 .... Long wheelbase cars introduced, version available with division between front and rear seats. Self-levelling height control deleted from front suspension. Air-conditioning standardised.

1970 .... Engine capacity increased from 6.25-litres to 6.75-litres. Central locking standardised.

1971 .... Mulliner, Park Ward two-door cars revised and renamed Corniche.

1972 .... 'Compliant' suspension introduced for all cars; radial-ply tyres specified.

1973 .... North American market cars fitted with energy-absorbing bumpers and pedal-operated parking brake; all cars have ventilated front disc brakes.

1974 .... Wheelbase extended, wider-section tyres fitted and wheelarches flared.

1975 .... Electronic ignition specified. Camargue, styled by Pininfarina, introduced complete with split level air conditioning.

1976 .... Corniche fitted with automatic air conditioning.

1977 .... Camargue and Corniche models uprated; rack and pinion steering. Silver Shadow II, T2 and Silver Wraith II introduced.

1979 .... Camargue and Corniche receive revised suspension. Standard Saloons continue.

1980 .... Bosch fuel injection introduced for Camargue and Corniche, California. Silver Shadow II, Silver Wraith II and T2 discontinued. Corniche Saloon discontinued.

1981 .... Fuel injection available for USA (except California - 1980). Corniche Convertible continues.

1985 .... Bentley Corniche renamed Continental. Camargue discontinued.

1986 .... Corniche II introduced for North America (rest of world had to wait until 1988).

1987 .... Fuel injection for all markets except California and rest of USA.

1988 .... Corniche II available for all markets - North America received it in 1986.

1990 .... Corniche III introduced.

1992 .... Corniche IV introduced.

1995 .... Corniche S and Bentley Continental Turbo available. Last Corniche and Bentley Continental produced.

Original specifications
Silver Shadow and T-Series Saloons

Coachwork

5-seat, 4-door Saloon of stressed steel monocoque construction. Separate front and rear sub-frames. Boot lid, doors and bonnet top of aluminium alloy. Individual and electrically-operated front seats; upholstery in English hide, deep pile carpets, washable headlining, walnut veneer facia and garnish rails. Padded top roll in black Ambla (PVC).

Dimensions

Wheelbase:	9ft 11.75ins (3035mm)
Track, front and rear:	4ft 9.5ins (1460mm)
Road clearance:	6.5ins (1650mm)
Turning circle:	38ft (11,580mm)
Overall length:	16ft 11.5ins (5170mm)
Overall width:	5ft 11ins (1800mm)
Overall height:	4ft 11.75ins (1520m)
Weight (unladen, complete with oil, coolant and a full tank of fuel):	4760lbs (2159kg)

Engine

8 cylinder vee unit with overhead valves and hydraulic tappets. Bore 4.1ins (104.14mm); stroke 3.6ins (91.44mm). 6230cc; compression ratio 9.0:1 (8.0:1 available). Cylinder block of high silicon content aluminium alloy with wet liners of cast iron; aluminium alloy cylinder heads; five bearing forged steel crankshaft. Full flow oil filter; water cooling; twin SU carburettors and automatic choke.

Transmission

Home market cars 4-speed automatic transmission; export models 3-speed GM400 gearbox. Ratios: (4-speed box) top: 1:1, 3rd: 1.45:1; 2nd: 2.63:1; 1st: 3.82:1, reverse: 4.3:1. (3-speed box) top: 1:1, 2nd: 1.5:1, 1st:2.5:1, reverse:2:1. Electric actuation fitted to both gearboxes.

Brakes

Disc brakes fitted to all wheels. Triple hydraulic system including two independent circuits. One power system provides 46 per cent of total braking, operating one caliper on each front disc and part of the rear brake. 2nd system provides 31 per cent total braking, operating the other caliper on each front disc. Direct master cylinder to part of the rear brakes provides 23 per cent of the total braking. Manually-operated parking brake.

Suspension

Independent all-round; double triangle levers, coil springs and telescopic hydraulic dampers at front; trailing arms, coil springs and telescopic hydraulic dampers at rear. Two-speed automatic height control; height sensor on front suspension and two sensors on the rear. Fast height control when doors open or neutral gear selected.

Steering

Power-assisted, Saginaw recirculating ball. 4 turns, lock-to-lock.

Wheels & tyres
15 inch, 5-stud, steel wheels, 8.45x15 low profile tyres. Tyre pressures: 23lbs/sq.in. front, 25lbs/sq.in. rear.

Electrical system
12-volt negative earth; 64 ampere-hour battery. Twin fuel pumps; rheostat on facia; 2-speed windscreen wipers; windscreen washers; electric fuel-filler cap release. Powered window lifts; electric seat adjustment; rear screen de-mister. Air conditioning optional.

Capacities
Fuel tank: .... 23.5 imp. galls., 28.8 USgalls., 109 litres. (Fuel tank later increased to 24 imp. galls.)
Engine oil: .... 14 imp. pints, 16.8 US pints, 8.0 litres.
Gearbox oil: .... 24 imp. pints, 28.8 US pints, 13.6 litres.
Coolant: .... 28 imp. pints, 33.6 US pints, 16 litres.

Colour schemes
Cars were finished in either a single colour or a two-tone scheme from the following range:
Black; Shell Grey; Tudor Grey; Black Pearl; Astral Blue; Sand; Sable; Sage Green; Smoke Green; Garnet; Dawn Blue; Regal Red; Velvet Green; Caribbean Blue.

Upholstery: Beige; Tan; Grey; Blue; Red; Green; Scarlet; Black.
Carpets: Beige; Blue; Fawn; Green; Grey; Maroon; Red.
Headlining: Grey; Light Fawn; Pale Green; Mushroom.

Performance
Maximum speed (average): ... 115mph, 185km/h 4390rpm
(best): 118mph, 190kph 4500rpm
3rd gear 72mph, 116kph 4000rpm
2nd gear 43mph, 69kph 4350rpm
1st gear 24mph, 39kph 3500rpm
Standing quarter mile: .... 17.6 seconds, 76mph, 121.60kph
Standing kilometre:.............. 33.0 seconds, 96mph, 153.60kph
Fuel consumption: .... 11-15mpg, 25.7 litres/100km - 18.8litres/100km; average 12.2mpg, 23.2 litres/100km

Derivative specification differed as follows:
Dimensions
Wheelbase (long): 10ft 3.5ins (3035mm)
Track:........................ Silver Shadow II & T2: front 5ft (1524mm), rear 4ft 11.5ins (1514mm); Camargue & Corniche: front & rear 4ft 9.5ins (1460mm).
Overall length: .... Long wheelbase saloon: 17ft 3.5ins (5270mm); Silver Shadow II & T2: 17ft 0.5ins (5194mm); Silver Wraith II: 17ft 4.5ins (5296mm); Corniche: 17ft '3.5ins (5270mm); Camargue: 16ft 11.5ins (51690mm); North American Silver Shadow II: 17ft 3.5ins (5270mm); North American Silver Wraith II: 17ft 8.5ins (5397.5mm).

Overall width: .... Silver Shadow II/T2: 5ft 11.75ins (1820mm); Corniche: 6ft (1830mm); Camargue: 6ft 3.5ins (1918mm)

Turning circle: .... Silver Shadow II, T2 & Silver Wraith II: 39ft 2ins (11,938mm); Corniche:38ft 9ins (11,811mm); Camargue: 38ft '6ins (11,735mm).

Weight: .... Silver Shadow II & T2: 4930lbs (2237kg); long wheelbase: 5010lbs (2275kg); Silver Wraith II: 5020 (2277kg); Silver Wraith II with division: 5260lbs (2385kg); two-door Saloon: 4978lbs (2258kg); two-door Convertible: 5124lbs (2322kg) ;Corniche Saloon: 5045lbs (2288kg); Corniche Convertible: 5200lbs (2358kg); Camargue: 5175lbs (2347kg).

6.75-litre engine
Stroke: 3.9ins (99.1mm)
Bore: 4.1ins (104.1mm)
Capacity: .... 6750cc.
Compression ratio:...... 9:1. From 1975, compression ratio 8:1 and 7.3:1 for American, Australian and Japanese markets.

Performance
Maximum speed: .... Silver Shadow II & T2: 119mph (190.4kph); Corniche: 120mph (192kph); Silver Wraith II: 119mph (190.4kph); Camargue:120mph (192kph).
Fuel consumption: ...... Silver Shadow II & T2: 13.6mpg (20 litres/100km); Silver Wraith II: 13.2mpg (22 litres/100km); Corniche: 11.9mpg (24 litres/100km); Camargue: 12.3mpg (23 litres/100km).

APPENDIX II
CLUBS, SPECIALISTS & BIBLIOGRAPHY

Clubs

Rolls-Royce Enthusiasts' Club
The Hunt House,
Paulerspury,
Towcester,
Northants. NN12 7NA,
England.
Tel: 01327 811788, Fax: 01327 811797.
The R-REC serves members throughout the United Kingdom, Europe and Canada. Organised sections exist in 19 areas within the UK, in 16 European countries and Canada. Members receive an excellent magazine, *The Bulletin*, six times a year as well as a monthly advertiser. The club was formed in 1957 and houses, at its headquarters, the entire build histories of Rolls-Royce and Bentley cars. Current membership is over 8000.

Rolls-Royce Owners' Club of Australia
Ian Dunn,
PO Box 163,
Lyneham,
ACT 2602.
Australia.
Tel: 0061 62545495
Formed in 1956, there are six sections throughout Australia. Magazine published six times a year.

Rolls-Royce Club of New Zealand
Tom Williams,
78 Kesteven Avenue,
Glendowie,
Auckland.
New Zealand.
The New Zealand Cub has three sections throughout the country. Meetings are arranged on a regular basis and members are kept informed of events through a magazine, which is published six times a year.

Rolls-Royce Owners' Club of America
Headquarters:
191 Hempt Road,
Mechanicsburg,
PA 17055.
USA.
Tel: 001 717 697 4671
The oldest Rolls-Royce Owners' Club in the world, the American club was formed in 1951. There are 33 regions throughout the USA and Canada and members receive the journal *Flying Lady* six times a year.

Bentley Drivers' Club
W.O. Bentley Memorial Building,
16 Chearsley Road,
Long Crendon,
Aylesbury,
Bucks HP18 9AW,
England.
Tel: 01844 208233.
The club has 8 regions in the United Kingdom and another 30-40 sections around the world. A quarterly magazine, the *Bentley Drivers Club Review*, is supplemented by a bi-monthly advertiser and a monthly newsletter. The club currently has a membership of around 3000.

Specialists

Rolls-Royce Motors Ltd.,
Pym's Lane,
Crewe,
Cheshire CW1 3PL,
England.
Tel: 01270 255155.

Rolls-Royce Motors Ltd.,
London Service Centre,
London,
England.
Tel: 0181 965 7355.

Official dealers:
P & A Wood,
Great Easton,
Dunmow,
Essex CM6 2HD,
England.
Tel: 01371 870848; Fax: 01371
870810.

Harwoods of Sussex,
London Road,
Pulborough,
Sussex,
England.
Tel: 01798 872407; Fax: 01798
872445.

Mann Egerton,
406 Derby Road,
Nottingham,
England.
Tel: 0115 978 0730.

Lancaster Europa Ltd.,
Hanover House, 2 Hanover Way,
Sheffield, S3 7UF,
South Yorkshire,
England.
Tel: 0114 276 5522.

Michael Powles Limited,
St. Peters Street,
Syston,
Leicestershire,
England.
Tel: 0116 260 1111.

Michael Powles Limited,
Duke Street,
Norwich,
Norfolk,
England.
Tel: 01603 219955.

Lancaster,
Kettering Road,
Northampton, NN1 4AJ,
England.
Tel: 01604 239944; Fax: 01604
234260.

Jacksons (Bournemouth) Ltd,
Holes Bay Road,
Poole,
Dorset BH15 2BD,
England.
Tel: 01202 666330; Fax: 01202
666130.

Weybridge Automobiles,
Brooklands Industrial Estate,
Vickers Drive,
Weybridge,
Surrey KT13 0YU,
England.
Tel: 01932 340231; Fax: 01932
354715.

Broughtons Limited,
Rutherford Way,
Cheltenham,
Gloucestershire GL51 9SQ,
England.
Tel: 01242 515374.

Reg Vardy plc.,
Tyne and Wear,
England.
Tel: 0191 5120101.

Murray Motor Company,
Bankhead Drive,
Sighthill,
Edinburgh,
Scotland.
Tel: 0131 442 2800.

Murray Motor Company,
North Street,
Charing Cross,
Glasgow,
Scotland.
Tel: 0141 221 6800.

Dutton Forshaw,
Riversway, Portway,
Ashton-On-Ribble,
Preston,
Lancashire,
England.
Tel: 01772 723456.

St. Helier Garages,
47 La Motte Street,
St. Helier, JE2 4SZ
Jersey, CI.
Tel: 01534 31341; Fax: 01534 23972.

Jack Barclay Limited,
Berkeley Square,
London.
Tel: 0171 629 7444;
Nine Elms,
London,
England.
Tel: 0171 738 8880.

H.R. Owen,
25 Old Brompton Road,
London SW7 3TD,
England. Tel: 0171 584 8451.

147

Riders of Falmouth,
Rider House,
Dracaena Avenue,
Falmouth,
Cornwall GR11 2EL,
England.
Tel: 01326 212222

Dutton-Forshaw,
Birtholt Road,
Parkwood,
Maidstone,
Kent NE15 9XL,
England.
Tel: 01622 692211.

Appleyard Rippon,
Roseville Road,
Leeds,
Yorkshire LS8 5QP,
England.
Tel: 0113 246 1111.

Paramount Cars Ltd,
Pentwin House,
Avenue Park,
Pentwin,
Cardiff CF2 7HE,
South Wales.
Tel: 01222 755766.

Olympic Ltd.,
Matford Park,
Marsh Barton,
Exeter,
Devon EX2 8FD,
England.
Tel: 01392 824515.

Mylchreest,
West Morland Road,
Douglas IM1 4AD,

Isle of Man.
Tel: 01624 623481.

John R Weir,
366 King Street,
Aberdeen AB24 5TR,
Scotland.
Tel: 01224 634211.

Henlys of Chester,
157-167 Foregate Street,
Chester,
Cheshire CH1 1HF,
England.
Tel: 01244 313901.

Stratstone of Wilmslow,
Altrincham Road,
Wilmslow,
Cheshire SK9 5NL,
England.
Tel: 01625 522222.

Mead of Burnham,
367 Bath Road,
Burnham, Slough,
Berkshire SL1 5QA.
Tel: 01628 668361.

Evans Halshaw Specialists Cars,
Monaco House,
Bristol Street,
Birmingham B5 7AU,
England.
Tel: 0121 666 6999.

Hadley Green Garage Ltd.,
202-204 High Street,
Barnet,
Hertfordshire EN5 5TA,
England.
Tel: 0181 440 8252.

Charles Hurst Ltd.,
62 Boucher Road,
Balmoral,
Belfast BT12 6LR,
Northern Ireland.
Tel: 01232 381721.

Maxwell Motors Ltd.,
Blackrock Bypass,
Blackrock
Co. Dublin,
Southern Ireland.
Tel: 00 3531 2885085.

Other specialists:
Coys of Kensington,
2-4 Queen Anne's gate Mews,
London SW7 5QJ,
England.
Tel: 0171 584 7444; Fax: 0171 584 2733.
Sales.

Brooks,
81 Westside,
London SW4 9AY,
England.
Tel: 0171 228 8000; Fax: 0171 585 0830.
Specialist auctioneers and valuers.

Hooper Alpe Limited,
50 Marylebone High Street,
London W1 3AD,
England.
Tel (Sales): 0171 935 1124; Fax: 0171 486 1488;
(Service): 0171 624 8833; Fax: 0171 328 8327.
German representative: Dr. Walter Leuthausel.
Tel: 0049 6403 71791;

Fax: 0049 6403 76032.
Sales and service.

Metex Autocovers,
Wood Street Mill,
Darwen
Lancashire BB3 1LS,
England.
Tel: 01254 704625/ 703893; Fax:
01254 776927.
Supplier of dust covers.

Frank Dale & Stepsons,
120-124 King Street,
Hammersmith,
London W6 0RH,
England.
Tel: 0181 748 0821; Fax: 0181 563
0359.
France, tel: (00 33 53 40 3000; Fax: 00
33 53 40 2420);
Germany, tel: (0049 211 404202, Fax:
0049 211 407764);
Japanese enquiries, tel: 0171 937 7432,
Fax: 0171 937 4828
Sales.

Sotheby's,
34-35 New Bond Street,
London W1A 2AA,
England.
Tel: 0171 314 4444; Fax: 0171 408
5909.
Auctioneers.

Exclusive Cars (Nottingham) Ltd.,
Unit 6, Pintail Close, Victoria Business
Park,
Netherfield,
Nottingham NG4 2PE,
England.
Tel: 0115 987 7277;

Fax: 0115 987 7600.
Sales and service.

Hillier Hill,
Unit 14, Stilebrook Road,
Yardley Road Industrial Estate,
Olney,
Bucks MK 46 5EA,
England.
Tel: 01234 713871.
Sales and service.

The Real Car Co.
Snowdonia Business Park,
Coed y Parc, Bethesda,
Gwynedd LL57 4YS,
Wales.
Tel: 01248 602649 (day), 01248 681572
(evening); Fax: 01248 600994.
Sales, service and spares.

The Chelsea Workshop,
Nell Gwynn House, Draycott Avenue,
Chelsea,
London SW3 3AU,
England.
Tel: 0171 584 8363-4/0171 581 1761/
0171 589 1522; Fax: 0171 581 3033.
Sales, service, restorations,
conversions.

County Radiators Ltd.,
32, Brook Road,
Rayleigh,
Essex,
England.
Tel: 01268 747001; Fax: 01268
745657.
Manufacturer and restorer of radiators
etc.

HighTone Restoration Ltd.,

Unit 5, Enstone Airfield,
Enstone,
Oxfordshire OX7 4NP,
England.
Tel: 01608 677328.
Restoration.

Creech Coachtrimming Centre,
45 Anerley Road,
Crystal Palace,
London SE19 2AS,
England.
Tel: 0181 659 4135.
Coachtrimming.

Brian Bilton-Sanderson F.I.M.I.
Maidenhead, Berkshire,
England.
Tel: 01628 74674.
Vehicle inspections.

Derby Plating Services Ltd.,
148 Abbey Street,
Derby,
England.
Tel: 01332 382408.
Plating specialists.

G. Whitehouse Autos Ltd.,
Haden Hill Road,
Halesowen,
West Midlands B63 3NE,
England.
Tel: 0121 550 7630; Fax: 0121 585
6408.
Transmission systems.

A.J. Hickman,
85 Worthington Road,
Fradley,
Lichfield, Staffordshire,
England.

Tel: 01543 252196.
Veneering and polishing.

Phantom Motor Cars,
Crondall, Farnham,
Surrey GU10 5QT,
England.
Tel: 01252 850 231; Fax: 01252 850516.
Sales & service, restoration, etc.

Classic Restorations, Scotland,
Pitnacree Street,
Alyth,
Perthshire PH11 8DY,
Scotland.
Tel: 01828 633293; Fax: 01828 632529.
Servicing and restoration.

Ashton Keynes Vintage Restorations Limited,
Ashton Keynes,
Swindon, Wiltshire,
England.
Tel: 01285 861288; Fax: 01285 860604.
Restoration, etc.

Prescote Motor Carriages,
Mill House,
Mill Road,
Totton,
Hampshire SO40 3ZQ,
England.
Tel: 01703 666682; Fax: 01703 66882.
Restoration, servicing, spare parts.

Servicentre Systems,
Somersham Road,
St. Ives
Cambridgeshire PE17 4LY,

England.
Exhaust systems.

Cover Systems,
113 High Street,
South Rushdcn,
Northants NN10 0RB,
England.
Tel: 01900 410851.
Manufacturer and supplier of car covers.

Caroline De Jonathan,
rue Libeau 36,
B-4682 Houtain Saint Siméon,
Belgium.
Tel: 0032 41 86 48 53/0032 41 86 48 54; Fax: 0032 41 86 48 56.
Restoration and conversion.

Peter Jarvis,
Gildenhill Place, Gildenhill Road,
Swanley,
Kent BR8 7PD,
England.
Tel: 01322 669081; Fax: 01322 662490.
Sales and spares.

Hoffmann's/Henley,
Faifield Works, Reading Road,
Henley-on-Thames,
Oxfordshire RG9 1DR,
England.
Tel: 01491 573953; Fax: 01491 573647.
Sales, service, etc.

Mike Thomas,
Linnet House, Lockgate Toad,
Sidlesham,
Chichester,

West Sussex PO20 7QQ,
England.
Tel: 01243 641007.
Coachtrimming.

London Chroming Company,
735 Old Kent Road,
London SE15,
England.
Tel: 0171 639 6434.
Plating and polishing.

Shadow Motorcars,
The Crest Complex, Courtenay Road,
Gillingham,
Kent ME8 0RX,
England.
Tel: 01634 264425.
Servicing and repairs.

David Beswick Coachtrimming,
18 Robinsons Lane Industrial Estate,
Shaftesbury Street,
Derby DE23 8NL,
England.
Tel: 01332 343252.
Coachtrimming.

Introcar Ltd.,
1 Manorgate Road,
Kingston-upon-Thames,
Surrey KT2 7AP,
England.
Tel: 0181 546 2027; Fax: 0181 546 5058.
New and used parts.

Silver Lady Motor Services Ltd.,
Hainult Works, Hainult Road,
Little Heath,
Romford,
Essex RM6 5SS,

England.
Tel: 0181 599 8548/4905; Fax: 0181 599 8041.
Service and repairs.

K.L.W.
137 Larkhall Lane,
Clapham, London SW4,
England.
Tel: 0171 622 8865/7473; Fax: 0171 978 1073.
Parts.

Ristes Motor Company Ltd.,
Gamble Street,
Nottingham,
England.
Tel: 0115 978 5834; Fax: 0115 9424351.
Steering column controls and switch plates.

Auto Interior and Hoods,
56 Norfolk Street,
Liverpool,
Cheshire L1 0BE,
England.
Tel: 0151 708 8881; Fax: 0151 708 6002.
Interior trim & carpets.

John Fletcher,
Rolls-Royce & Bentley Brokerage.
Tel: 01678 520321; Fax: 01678 521234.
Brokerage.

Garage De Vaal,
Heulweg 78,
2295 KH Kwintsheul,
The Netherlands.
Tel: 0031-174-510022/297545; Fax:

0031-174-298175.
Rolls-Royce and Bentley specialist.

Woodlift Limited,
Woodlift House, 106 Roebuck Road,
Chessington Industrial Estate,
Chessington,
Surrey KT9 1EU,
England.
Tel: 0181 974 1934; Fax: 0181 974 1935.
Woodwork specialist.

Sebright Garage,
Barnet, Hertfordshire,
England.
Tel: 0181 440 3115.
Service and repairs.

Hanwell Car Centre,
Broadway, 86/88 Uxbridge Road,
Hanwell, London W7,
England.
Tel: 0181 567 6557/9729; Fax: 0181 579 5386.
Sales.

John Stuart Dennison,
54 Broadbent Road,
Watersheddngs,
Oldham,
Lancashire OL1 4HY,
England.
Tel: 0161 652 4544.
Sales and R-R/Bentley memorabilia.

Gary Bretherton,Motor Engineer,
Unit J, Grove Mill, Grove Crescent,
Eccleston, Chorley,
Lancashire PR6 9RS,
England.
Tel: 01257 453531. Servicing.

Bowling-Ryan Ltd,
Unit 5 Fishbrook Industrial Estate,
Stoneclough Road,
Kearsley, Bolton,
Lancashire BL4 8EL,
England.
Tel: 01204 700300.
Servicing and restoration.

Brunts of Silverdale,
Stonewall,
Silverdale, Newcastle
Saffordshire ST5 6NR,
England.
Tel: 01782 625225; Fax: 01782 717530.
Servicing and restoration.

Harvey Bailey Engineering (HBE),
Ladycroft Farm,
Kniveton, Ashbourne,
Derbyshire DE6 1JH,
England.
Tel: 01335 346419; Fax: 01335 346440. Suspension and handling kits.

Rob Jones, Benver Services,
Unit 9, Quaker Coppice, Crewe Gates
Industrial Estate,
Crewe,
Cheshire CW1 6FA,
England.
Tel: 01270 250236.
Servicing and restoration.

P.J. Fischer Classic Automobiles,
Northumberland Garage,
Dyers Lane, Upper Richmond Road,
Putney, London SW15,
England.
Tel: 0181 785 6633; Fax: 0181 785 6926. Sales.

Ivor Bleaney of the New Forest,
PO Box 60,
Salisbury,
Wiltshire SP5 2DH,
England.
Tel: 01794 390895;
Fax: 01794 390862. Sales.

Michael Hibberd,
Unit 31 Middle Green Trading Estate,
Middle Green, Langley,
Slough,
Berkshire SL3 6DF,
England.
Tel: 01753 531631.
Service, restoration and sales.

Best Of British,
Hobbemaweg 90,
6562 CV Groesbeek,
Holland.
Tel: 0024 3977391; Fax: 0024 3977244.
Sales.

Max of Switzerland,
6913 East McDowell Rd.,
Scottsdale 85257, Arizona,
USA.
Tel: 001 602 945 4545.
Sales & service.

Matthews Motor Co.,
4901 North Oracle Rd.,
PO Box 27878, Tuscan, Arizona,
USA.
Tel: 001 602 888 7900.
Sales & service.

Rolls-Royce of Beverley Hills Ltd.,
9018, Wilshire Boulvevard,
Beverley Hills, 90211, California,

USA.
Tel: 001 213 659 4050.
Sales & service.

Terry York Motor Cars,
15800 Ventura Boulevard,
Enrico 91436, California,
USA.
Tel: 001 213 990 9870.
Sales & service.

British Motor Cars of Monterey Inc.,
777 Del Monte Ave.,
Monterey, 93940, California,
USA.
Tel: 408 373 3041.
Sales & service.

Roy Carver Inc.,
1540 Jamboree Rd.,
Newport Beach, 92660-0180,
California,
USA.
Tel: 001 714 640 6444.
Sales & service.

Peter Epsteen Ltd.,
68-131 Highway 111,
Palm Springs 92262, California,
USA.
Tel: 001 714 328 8981.
Parts.

PR Parts Inc.,
PO Box 4462, Glendale, CA 91202,
USA.
Tel: 001213 500 7600.
Parts.

Tony Handler,
2028 Cotner Ave.,
Westwood, CA 90025,

USA.
Tel: 001 213 473 7773.
Parts, new and used.

The Brassworks,
289 Prado Road,
San Luis Obispo, California 93401,
USA.
Tel: 001 342 6759; 544 8841; Fax: 001 544 5615.
Hand-crafted radiators.

Peter Satori Co. Ltd.,
285-325 West Colorado Boulevard,
Pasedena 91105, California,
USA.
Tel: 001 213 681 8123.
Sales & service.

British Motor Car Distributors Ltd.,
901 Van Ness Ave.,
San Francisco 94109, California,
USA.
Tel:001 415 776 7700.
Sales & service.

Imported Cars of Greenwich Inc.,
217 West Putnam Ave.,
Greenwich, 06830, Connecticut,
USA.
Tel: 001 203 869 2850.
Sales & service.

Lauderdale Motor Car Corp.,
407 North Federal Highway,
Fort Lauderdale 33301, Florida,
USA.
Tel: 001 305 764 5881.
Sales & service.

Val Ward Imports Inc.,
8700 S.Tamiiami Trail,

Fort Myers, 33907, Florida,
USA.
Tel: 001 813 939 4616.
Sales & service.

Gregg Motor Cars Inc.,
10231 Atlantic Boulevard,
Jacksonville, 32211, Florida,
USA.
Tel: 001 904 724 1080.
Sales & service.

Braman Motors Inc.,
2020 Biscayne Boulevard,
Miami 33137, Florida,
USA.
Tel: 001 305 576 6900.
Sales & service.

Scarritt Motors Inc.,
555, 34th St. S.
St. Petersburg, 33711, Florida,
USA.
Tel: 001 813 327 3700.
Sales & service.

Royal Motorcar Corp.,
1314 South Dixie Highway,
West Palm Beach,33401, Florida,
USA.
Tel: 001 305 659 1314.
Sales & service.

Vantage Motorworks, Inc.,
1898 N.E. 151 St.,
Miami, Florida 33162,
USA.
Tel: 001 305 940 1161.
Sales, service, restoration.

Classic Auto Rstorations,
22456 Orchard Lake Rd.,

Farmington, MI 48024,
USA.
Tel: 001 313 477 4767.
Parts.

Proper Moor Cars Inc.,
1811-11th Avenue North,
Saint Petersburg, Florida 33713,
USA.
Tel: 001 813 821 8883.
Sales, service, restoration.

R&B Parts,
4546 Palm Beach Canal Rd.,
West Palm Beach, FL 33406,
USA.
Tel: 001 305 689 7888
Parts.

Mitchell Motors Inc.,
5675 Peachtree Industrial Boulevard,
Chamblee 30341,
Georgia,
USA.
Tel: 001 404 458 5111.
Sales & service.

Continental Cars Ltd.,
1072 Young St.,
Honolulu 96814, Hawaii,
USA.
Tel: 001 808 526 3258.
Sales & service.

Worden-Martin Inc.,
100 Carriage Center,
2003 S. Neil St.,
Champaign 61820,
Illinois,
USA.
Tel: 001 217 352 7901.
Sales & Service.

Loeber Importers Ltd.,
5625 North Broadway,
Chicago 60660, Illinois,
USA.
Tel: 001 312 728 5000.
also at:
1111 North Clark St., Chicago 60610,
USA.
Tel: 001 312 944 0500.
Sales & service.

Continental Motors Inc.,
420 E Ogden Ave.,
Hinsdale, 60521, Illinois,
USA.
Tel: 001 312 655 3535.
Sales & service.

Dave Lewis Restoration,
3825 South Street, Second Street,
Springfield, Illinois 62703,
USA.
Tel: 001 217 529 5290.
Restoration.

Albers Rolls-Royce,
360 S. First St.,
Zionsville, IN 46077,
USA.
Tel: 001 317 873 2360.
Parts.
Euro Motor Cars Bertseda Inc.,
4800 Elms St.
Berthseda, 0814, Maryland,
USA.
Tel: 001 301 986 8800.
Sales & service.

Foreign Motors West,
253 N. Main St.,
Natick, MA 01760,
USA.

Tel: 001 617 653 4323.
Parts.

LeBaron Bonney Company,
Dept. 212, PO Box 6 Chestnut Street,
Amesbury, Massachcsctts 01903,
USA.
Tel: 001 508 388 3811; Fax: 508 388
1113.
Upholstery fabrics.

Sears Imported Autos Inc.,
13500 Wayzata Boulevard,
Minnetonka, 55343, Minnesota,
USA.
Tel: 001 612 546 5301.
Sales & service.

Charles Schmitt & Co.,
3500 South Kingshighway Boulevard,
St.Louis 63139, Missouri,
USA.
Tel: 001 314 352 9100.
Sales & service.

Cutter Motorcars,
2333 South Decatur Boulevard,
Las vegas 89102, Nevada,
USA.
Tel: 001 702 871 1010.
Sales & service.

Modern Classic Motors,
3225 Mill St.,
Reno 89501, Nevada,
USA.
Tel: 001 702 323 4169.
Sales & sevice.

Imported Motor Car Co.
34 Valley Rd.,
Montclair 07042, NJ,

USA.
Tel: 001 201 746 4500.
Sales & service.

Turner Spares Ltd.,
Raritan Center Parkway,
Box 396 New Jersey 08818,
USA.
Tel: 001 201 225 5800.
Spares.

Knight-Clarke Services,
8 Bodine Ave.,
Gladstone, New Jersey.
Tel: 201 234 2930.
Parts.

Perfection Motor Car Co.,
6012 Acadamy Rd. N.E.,
Albuquerque, 87109, New Mexico,
USA.
Tel: 001 505 822 8500.
Sales & service.

Rallye Motors Inc.,
20 Cedar Swamp Rd.,
Glen Cove, 11542, NY,
USA.
Tel: 001 516 671 4622.
Sales & service.

George Haug Co. Inc.,
517 East 73rd St. New York 10021,
USA.
Tel: 001 212 288 0173.
Service.

Park Ward Motors Inc.,
301 East 57th St., New York 10022,
USA.
Tel: 001 212 688 7112.
Sales & sevice.

Premier Resource Group Inc.,
New York,
USA.
Tel: 001 212 730 5823; Fax: 001 212
354 1323
Sales.

The Belmont Group,
USA.
Tel: 001 708 945 9603; Fax: 001 708
945 9636.
Sale & purchase of Rolls-Royce and
Bentley cars anywhere in world.

Bob's Auto Parts,,
Rt. 9W,
Kingston, NY 12401,
 USA.
Tel: 001 914 336 6330.
Used parts.

Transco Inc.,
1800 N.Main St.,
High Point 27262,
North Carolina,
USA.
Tel: 001 919 885 5171.
Sales & service.

John McCombie Inc.,
572 S. Nelson Rd.,
Columbus, Ohio 43205,
USA.
Tel: 001 614 221 2563.
Parts.

Jackie Cooper Imports Inc.,
9505 North May Ave.,
Oklahoma, 73120,
USA.
Tel: 001 405 755 3600.
Sales & service.

Siggi Grimm Inc.,
2007 East 11th St.,
Tulsa, 74104, Oklahoma,
USA.
Tel: 001 918 582 1151.
Sales & service.

Mente Shelton Motor Co.,
1638 West Burnside St.,
Portland, 97228, Oregon,
USA.
Tel: 001 503 224 3232.
Sales & service.

Keenan Motors Inc.,
3900 Broad St.,
Philadephia, 19140, Pennsylvania,
USA.
Tel: 001 215 223 4600.
Sales & service.

Ascot Imported Cars Inc.,
418 Walnut St.,
Sewickly, 15143, Pennsylvania,
USA.
Tel: 001 412 761 9310.
Sales & service.

Dick Dyer Assoc.,
5717, Two Notch Rd.,
Columbia, 29204, S.Carolina,
USA.
Tel: 001 803 786 2010.
Sales & service.

Autorama Inc.,
2950 Airways Boulevard,
Memphis, 38130, Tennessee,
USA.
Tel: 001 901 345 6211.
Sales & service.

Superior Motors Inc.
630 Murfreesboro.Rd.,
Nashville, 37210, Tennessee,
USA.
Tel: 001 615 254 5641.
Sales & service.

Overseas Motors Of Dallas,
7018 Lemmon Ave.,
Dallas, 75209, Texas,
USA.
Tel: 001 214 358 1446.
Sales & service.

Overseas Motors Corp. Of Fort Worth,
2824 White Settlement Rd.,
Fort Worth,76107, Texas,
USA.
Tel: 001 817 332 4181.
Sales & service.

Ken Garff Imports Inc.,
525 Soth State St.
Salt Lake City, 84111, Utah,
USA.
Tel: 001 801 521 6604.
Sales & service.

Dominion Rolls-Royce Ltd.,
6517 West Broad St., Richmond,
23230, Virginia, USA.
Tel: 001 804 288 3171.
Sales & service.

Uptown Motors Inc.,
2111 North Mayfair Rd.,
Milwaukee, 53226, Winconsin,
USA.
Tel: 001 414 771 9000.
Sales & service.

Bibliography

The Rolls-Royce Motor Car Anthony Bird & Ian Hallows (Batsford)

Rolls-Royce and Bentley Klaus-Josef Rossfelt (Haynes ISBN 0-85429 920 3)

Rolls-Royce and Bentley, The Crewe Years Martin Bennett (Haynes ISBN 0 85429 908 4)

Rolls-Royce Bentley Experimental Cars Ian Rimmer (R-REC ISBN 1 869912 00 4)

The Rolls Royce and Bentley Vols 1, 2 & 3 Graham Robson (Motor Racing Publications ISBN 0 900549 86 6; 0 900549 87 4; 0 900549 99 8)

Those Elegant Rolls-Royce Lawrence Dalton (Dalton Watson Ltd)

Rolls-Royce In America John Webb de Campi (Dalton Watson Ltd)

Rolls-Royce, The Living Legend (Post Motor Books)

Rolls-Royce, 80 years of Motoring Excellence Edward Eves (Orbis Books ISBN 0 85613 647 6)

Rolls-Royce (Autocar archives) (Temple Press ISBN 0 600 34981 0)

The Rolls-Royce Jonathan Wood (Shire Publications ISBN 0 85263 873 6)

The Bentley Nick Georgano (Shire Publications ISBN 0 7478 0192 4)

The Life Of Sir Henry Royce Sir Max Pemberton (Huchinson)

An Illustrated History Of The Bentley Car W.O. Bentley (George Allen & Unwin Ltd)

Bentley Past & Present Rivers Fletcher (Gentry Books ISBN 085614 082 1)

Bentley, The Cars From Crewe Rodney Steel (Dalton Watson ISBN 0 901564 45 1)

Rolls-Royces - Hive's Turbulent Barons Alec Harvey-Bailey (Sir Henry Royce Memorial Foundation)

Bentley R-Type Continental Stanley Sedgwick (Bentley Drivers' Club Ltd)

Illustrated Rolls-Royce & Bentley Buyer's Guide Paul R. Woudenberg (Motorbooks International)

Postwar Rolls-Royce and Bentley, A Concise Buying Guide Barry D.Cooney (Cooney-Taylor Publishing Inc ISBN 0 916117 00 6)

Bentley Heritage Richard Bird (Osprey Automotive ISBN 1 85532 187 4)

Bentley Cars 1940-1945 (Brooklands Books)

Rolls-Royce Silver Shadow Gold Portfolio, 1965-1980 (Brooklands Books ISBN 1 85520 2298)

Road & Track *on Rolls-Royce and Bentley* (Brooklands Books ISBN 0 946489 59 9)

Rolls-Royce Silver Shadow 1 & II (Transport Source Books ISBN 1 85847 298 9)

Rolls-Royce Corniche, Camargue & Silver Spirit (Transport Source Books ISBN 1 85847 299 7)

Rolls-Royce Silver Dawn, Cloud & Phantom (Transport Source Books ISBN 1 85847 281 4)

From The Shadow's Corner Cal West (Rolls-Royce Owners' Club Inc)

The Motor Car 1946-56 Michael Sedgwick (Batsford ISBN 0 7134 1271 2)

Britain's Motor Industry, The First Hundred Years Georgano, Baldwin,Clausager & Wood (Haynes ISBN 0 85429 923 8)

British Car Factories From 1896 Collins & Stratton (Veloce Publishing ISBN 1 874105 04 9)

Rolls-Royce Enthusiasts' Club *Bulletin*

Motor

Autocar

Autosport

Motor Sport

Classic Cars

Queste Magazine

INDEX

Dear Reader,
We hope you have enjoyed this
Veloce Publishing production. If
you have ideas for books on Rolls-
Royce or Bentley, or other
marques, please write and tell us.
Meantime, Happy Motoring!